THE FEATHERED ONION

D1016873

THE FEATHERED ONION

CREATION OF LIFE IN THE UNIVERSE

Clive Trotman

WILEY

Published in 2004 by   John Wiley & Sons, Ltd, The Atrium, Southern Gate
Chichester, West Sussex, PO19 8SQ, England
Phone   (+44) 1243 779777

Copyright © 2004   Clive Trotman

Email (for orders and customer service enquiries): cs-books@wiley.co.uk
Visit our Home Page on www.wiley.co.uk or www.wiley.com

All Rights Reserved. No part of this publication may be reproduced, stored in a retrieval system or
transmitted in any form or by any means, electronic, mechanical, photocopying, recording,
scanning or otherwise, except under the terms of the Copyright, Designs and Patents Act 1988 or
under the terms of a licence issued by the Copyright Licensing Agency Ltd, 90 Tottenham Court
Road, London W1T 4LP, UK, without the permission in writing of the Publisher. Requests to the
Publisher should be addressed to the Permissions Department, John Wiley & Sons Ltd, The
Atrium, Southern Gate, Chichester, West Sussex PO19 8SQ, England, or emailed to
permreq@wiley.co.uk, or faxed to (+44) 1243 770620.

Designations used by companies to distinguish their products are often claimed as trademarks.
All brand names and product names used in this book are trade names, service marks, trademarks
or registered trademarks of their respective owners. The Publisher is not associated with any
product or vendor mentioned in this book.

This publication is designed to provide accurate and authoritative information in regard to the
subject matter covered. It is sold on the understanding that the Publisher is not engaged in
rendering professional services. If professional advice or other expert assistance is required, the
services of a competent professional should be sought.

Clive Trotman has asserted his right under the Copyright, Designs and Patents Act 1988, to be
identified as the author of this work.

Other Wiley Editorial Offices

John Wiley & Sons, Inc. 111 River Street, Hoboken, NJ 07030, USA

Jossey-Bass, 989 Market Street, San Francisco, CA 94103–1741, USA

Wiley-VCH Verlag GmbH, Pappellaee 3, D-69469 Weinheim, Germany

John Wiley & Sons Australia, Ltd, 33 Park Road, Milton, Queensland, 4064, Australia

John Wiley & Sons (Asia) Pte Ltd, 2 Clementi Loop #02–01, Jin Xing Distripark, Singapore 129809

John Wiley & Sons Canada Ltd, 22 Worcester Road, Etobicoke, Ontario, Canada, M9W 1L1

Wiley also publishes its books in a variety of electronic formats. Some content that appears in
print may not be available in electronic books.

Library of Congress Cataloging-in-Publication Data

British Library Cataloguing in Publication Data
A catalogue record for this book is available from the British Library

ISBN 0–470–87187–3

Typeset in Melior $9\frac{1}{2}$/13pt by Mathematical Composition Setters Ltd, Salisbury, Wiltshire
Printed and bound in Great Britain by T.J. International

This book is printed on acid-free paper responsibly manufactured from sustainable forestry in
which at least two trees are planted for each one used for paper production.

10 9 8 7 6 5 4 3 2 1

*To Jeanette,
Paul, Andrew,
Saxon, Campbell, Caitlyn*

Contents

Preface xi

Chapter 1 Year Dot 1

The Earth is very old, the universe is older still. Life had plenty of time to develop
How much time has been available on the Earth, and in the universe, for life to form? Methods of estimation, early and more recent. The Earth is now dated 4.6 billion years by radioactivity; the universe is dated 10–20 billion years by observations of star light.

Chapter 2 Friends and Relatives 30

All life on Earth is related to a single ancestor
All forms of life are much more closely related than first appearances might suggest. Major differences, even between animals, plants and bacteria, are superficial. The inner workings of their cells are virtually identical. Nearly all life depends on the same source of energy, which is the simple combination of hydrogen and oxygen to make water. Life on Earth had a single origin, making the search for it easier.

Chapter 3 Dating the Ancestors 53

When a timescale is added to relatedness, life is found to be very old
Historical ideas about the youth of the Earth, some of the earlier beliefs and influential personalities. Recent progress in dating fossils. Scientific distortions such as hypothetical 'missing links' and the Piltdown Man.

Chapter 4 Before the Ancestors   70
Life is at least as old as the Earth
New technology enables protein or DNA sequences to be
compared, but a fresh argument questions whether this
provides a reliable evolutionary timescale.

Chapter 5 Life's Not Simple   91
Life on Earth has always been complex
'Primitive' life more than 3.8 billion years ago was already
highly complex, with cells, genes, proteins and an
intricate biochemical metabolism.

Chapter 6 Thanks to Thermodynamics   110
If life was never simple, how did it start?
The central paradox of life: since life can only be complex,
how can it ever have been simple? The evolution of life's
chemistry happened in the 10 billion years or so before
the Earth existed.

Chapter 7 Non-Event   133
The moment life did not come into existence
There was no specific event.

Chapter 8 Spreading the Message   149
Life is universal – but don't bother searching for it
Doubling processes, such as gene duplication and cell
division, are so fundamental to life that a single primitive
cell, almost regardless of its inefficiency, could colonise a
sterile ocean in a blink of geological time. Ice comets
could preserve and transport inter-stellar chemistry. The
Oort Cloud and Kuiper Belt are great reservoirs of
cometary material that can survive passage through the
atmosphere into the oceans of the Earth.

Chapter 9 Unintelligent Design 177
Life's inheritance
Life's timescale is at least that of the universe, not merely the Earth. Life has changed very little in the Earth's accepted timescale of 4.6 billion years. Evolution has been merely a few simple variations on an underlying biochemical theme. Innovations have been trivial. Far from the age of the Earth providing any constraint on the antiquity of life, ultimately an understanding of the origin of life may throw fresh light on the historical timeframe of the universe.

Chapter 10 Life: To Be Continued? 197
Life could do better, but probably won't
A genome is a program for the construction of a living being and we are on the point of being able to rewrite that program to manufacture any grotesque combination. The human species has reached the critical point where it can change its own destiny. On the other hand, human intelligence and social behaviour have changed little in thousands of years and will change little in future millennia.

Glossary 225

Appendix 232
Simple calculations, sources of information and wider reading.

Index 249

Where the telescope ends, the microscope begins. Which of the two has the greater view?

Victor Hugo (1802–1885),
Les Misérables, 1862

How often have I said to you that when you have eliminated the impossible, whatever remains, *however improbable*, must be the truth?

Sir Arthur Conan Doyle (1859–1930),
The Sign of Four, 1890

...your judgment will probably be right, but your reasons will certainly be wrong.

Lord Mansfield (1705–1793),
Chief Justice of England 1756–1788

PREFACE

Could life have originated from scratch on planet Earth or did it begin elsewhere? Life as we know it today is obviously complex, yet its earliest traces on Earth had numerous features that even now seem highly advanced. Although the Earth is very ancient its beginnings can be dated quite accurately. Life is also very ancient, but how ancient is much less certain. And that's the problem. If life turns out to be considerably older than the Earth, then there's a lot of explaining to do.

The exploration of this problem begins nevertheless with a certainty: life has existed for billions of years. That's not seriously in dispute, but whether it means 3 or 4 billion years (less than the age of the Earth) or perhaps 6 or even 12 billion years (considerably longer than the Earth has existed, approaching the age of the universe) – well, that's the question. Life just might be sufficiently young to have started here after the Earth came into existence 4.6 billion years ago. But if life is older, it must have started at an unknown place at an unknown time, or places and times, elsewhere in the universe.

Where and when, exactly? Nowhere; everywhere, all the time. No particular place or time needs to be pinned down. As soon as the new universe was cool enough for molecules to be built from atoms, the path towards life was inevitable. The components were forming everywhere and the laws of thermodynamics stirred the mix to ensure that they eventually came together.

As little as 150 years ago the vast majority of people believed that life, complete with all the species, had been created just the way it is, and created quite recently. Before evolution was discovered, an origin for life was not even a question. The belief that the world and all its flora and fauna were created especially for humans was very comforting, but was not the result of rational enquiry.

This is not a book on evolution, yet to some extent it concerns evolution, because that reality had to be accepted before people were prepared to go further and think about an origin of life. Much of the evidence for evolution was easily observed and quite well known hundreds of years ago, but the universal mindset was to push the evidence out of sight and out of mind. Extinct fossils were dismissed with every contrived explanation except one: that they were extinct fossils. Few people accepted the sweepingly logical deduction that complex life developed from simple life over a very long period of time. But how much further forward are we today, a century and a half after Darwin's *Origin of Species*? Not a lot: surveys indicate that a surprisingly large proportion of people, perhaps around half, are not ready to accept the ideas that life created itself spontaneously and evolved naturally over time.

Over the past century or so, two important dates have been pushed steadily further into the remote past. One is the date of the Earth's formation. An increasingly ancient Earth has been comforting in providing more time for evolutionary events to have occurred – up to a point. But one problem has been exchanged for another, because the other date to have been revised progressively earlier is the first appearance of life on Earth. Innovations in science and technology have opened up a startlingly clear view of ancient life that now converges in time on the formation of the Earth itself. As the origin of life has slipped progressively earlier, the interval between the formation of a hospitable surface on the Earth and the appearance of primitive life has effectively disappeared. There just wasn't enough time for life to come into existence.

Something has to give. Is the Earth older than we think? Did life on Earth originate incredibly rapidly? Did life or its components originate somewhere else before finding their way to our planet?

The trail has been well obscured. The earliest species have disappeared altogether and been replaced many times over. Hardly any living organisms are a century old, yet life has existed on our planet for about 40 million centuries. Regardless of how 'primitive' some modern life forms appear to be, they are not in fact primitive at all, having been refining themselves, through their antecedents and the process of selection, for billions of years. All life on Earth today is the same age – antiquity might be a better word – having descended from the first enduring germination of life.

Most of the vestiges left by primordial life were wiped out long ago – but not all. The clues left behind are often microscopic and many are as small as atoms, but they live on, and by asking the correct questions and applying a little inescapable logic we can edge just a little closer to the facts. For example, why do the carbon atoms found in some extremely ancient sediments appear to have been sorted by mass in a way that only life sorts them – are they first-hand witnesses of primordial life? Why do un-questionably ancient oils have biochemical signatures uncannily similar to modern life? Where did the atmosphere obtain the oxygen on which animal life so depends for respiration? If the answer is from plants, that throws up a new problem because plants both photosynthesise (harness the Sun's rays for synthesis) and respire (burn fuel and oxygen for energy) whereas animals do only the latter. Since this makes plants more complex than animals, not more simple, how come plants appeared first? The familiar red haemoglobin of blood carries oxygen around the body, but molecular clues show that haemoglobin existed long before multicellular bodies did, and it existed in plants. Whatever for?

These clues, coupled with the powerful scientific principle that the general case (life could have originated anywhere, anytime) is always a sounder proposition than special pleading (Earth is unique in having life), lead to serious questions as to whether life evolved from scratch on this Earth.

Good scientific thinking is never to assume more than the minimum necessary to explain the observations. The sure sign of a good scientific theory is a sweeping simplicity without the need for special conditions to explain away any awkward bits. $E = mc^2$; what can anyone add? The simplest view of life is that at one time there wasn't any, and now there is. Some time in between, it appeared. Logic demands that life progressed from simple to complex and not the other way about. The conundrum, that the earliest primitive life seems actually to have been highly complex, can lead to the trap of so-called irreducible complexity, the easy assumption that a certain level of life (in effect, a cell) is the minimum below which life could not exist, because taking away any of its component parts would be fatal. 'Irreducible complexity' is an entirely artificial paradox that doesn't really exist and can readily be resolved, a figment of an arbitrary definition of life. The complexity of early life is not at all irreducible – but it is irrefutable.

In the past 30 years new technologies have yielded a rich harvest of molecular data providing fresh evidence of life's origins. The DNA of genes contains the information to duplicate life (but, interestingly, not to make life). Most of this information is expressed as the structure of proteins. Closely related species have closely related proteins. Distantly related species have more distantly related proteins, but they are nonetheless still related. Modern molecular technology makes it possible to measure protein relationships, to put numbers and ages to them. This is where the fun starts, because it soon becomes obvious that modern, complex proteins have exceedingly ancient histories. It follows that primordial life on Earth was capable of mass-producing highly complex proteins, a process demanding extremely intricate biochemistry.

It is easy to be misled by the superficial complexity of supposedly advanced life such as humans and flying insects, porpoises and fly-catching plants, but beneath the surface their cells and their biochemistry are remarkably similar. Their complexity lies not in their antics but right inside their cells. As powerful microscopical techniques have been teamed with ways of visualising molecules atom by atom, as the internal workings of membranes have been teased out and reconstructed with computer models, as the language of the genes has been translated, it has become clear that the structural complexity of each individual cell is almost beyond comprehension. Nano-sized molecular motors spin inside cells like invisibly tiny turbines. In one respect complexity is irreducible – these features are likely to have been similarly complex billions of years ago when primitive life consisted only of single cells but, as we shall see, functioned like modern cells. However, if biochemistry has changed little in the past three or four billion years, then, inescapably, early life was virtually as complex as it is now. If life did not achieve such stunning complexity quickly, the process must have begun before the timeframe of Earth's existence. Our overall task is to assess the likelihood of this being the case and look for clues to the ways and means.

As the story unfolds, it becomes clear that any attempt to predict life as it will be in 10 or 100 million years' time is futile, beyond the conclusion that at the cellular level it may never become very different. Nevertheless at the macroscopic level, the whole body, the opposite is true: the lesson of life's history is that nothing about bacteria or any other single-celled life could have prompted an observer of primitive Earth to predict the evolution of trees or bees, feathers or onions. Whatever life turns out to be like in the far distant future, the one certainty is that nobody alive today has any chance of predicting it correctly.

Of course, many people for varied reasons will tell me that I am wrong about life's origins. Nothing would give me greater pleasure than to be *proved* wrong, because that is precisely how science makes progress. Science does not recognise proof of being right, only disproof. There is all the difference between disproof and disbelief, however, and I have no wish to change anyone's beliefs. It is a sad fact of life that the less a subject is understood the more vehemently people defend their own ideas, but in this debate we should all be on the same side: the search for rational explanation. As Spanish philosopher Miguel de Unamuno (1864–1936) commented, 'True science teaches, above all, to doubt and to be ignorant.'

It would be impossible to satisfy every reader in a book of this kind. I have chosen a readership with a broad interest in science, but not an academic readership. It simply is not possible to incorporate in one work complete textbooks of animal and plant biology, chemistry, nuclear physics, geology, genetics, molecular biology, palaeontology and statistics! For readers who demand chapter and verse on sweeping statements and calculations, some background information and references are provided in the Appendix. Most of the references can be found in a reasonable public library (or the library can easily borrow them), and these lead to more specialised references that the enthusiast can pursue if desired.

Many people have kindly explained to me things outside my field so that I can attempt to explain them to you. The advent of the Internet has made it possible to contact expert sources of information (and disinformation) all over the world. What is remarkable is that complete strangers who have never met, and probably never will, are happy to devote time and energy to the exchange of scientific ideas and information over the Internet. I thank all the helpful people I have met in this way.

The recurring nightmare when planning a research lecture is to project the customary slide acknowledging all the people to be credited, then to realise with horror that a person has been left off – predictably the most crucial and easily wounded. After much agonising I decided to delete the list drawn up for this book, not only fearing the risk of omission but also not wanting to imply that anyone's help meant that they agreed with my conclusions. Most were highly qualified professionals and I will not enhance my own credibility through other people's credentials. Similarly, acknowledgements to people named in the text or listed in the references do not in any way imply that they would agree with my interpretations. To everyone who has helped me, often through spirited argument, I offer my personal and sincere thanks.

There are two exceptions who must be singled out, however. They are (chronologically) my extremely hard-working and assiduous literary agent, Lavinia Trevor of London, who never misses a thing on behalf of her clients. And Sally M. Smith, publisher, of Wiley, for her breathtaking ability to place herself in the shoes of precisely the readership I had in mind, always asking the most penetrating questions and never failing to spot the points, large or small, that I had so often fudged. All errors, of course, remain my own responsibility.

1

YEAR DOT

The Earth is very old, the universe is older still.
Life had plenty of time to develop

A biologist, entertaining friends and relatives at a restaurant, attracted the waiter's attention. 'That,' he said disdainfully, handing the waiter a soggy neck bone from what was supposed to be chicken casserole, 'is not from any bird. That's a mammalian bone.' The bone, as the biologist and the restaurant proprietor knew perfectly well, was rabbit.

The wine was with the compliments of the house! But the real point is that it is remarkably difficult for the non-expert to distinguish between a mammal and a bird stripped of fur or feathers. Despite the unmistakable differences between a two-legged bird with wings and a beak and a four-legged rabbit with fur and long teeth, it takes some degree of expertise, or taste, to say whether a piece from the inside belongs to one or the other.

Can rabbits, birds and indeed humans be so closely related that they all had a single common ancestor hundreds of millions of years ago? To many people this is simply not believable, yet the proof is indisputable and becoming steadily firmer. More difficult still to accept is that humans, birds, snails and plants can be descended from an ancestor that lived not just hundreds of millions but billions of years ago. Nevertheless that's the end of the story, not the beginning. Not only is life today very complex, life on Earth has always been complex. Science is about measurement and our starting point is to find out the age of the Earth and the age of the universe, and thus to provide a timeframe into which to fit the antiquity of life – whatever that may turn out to be. The Earth and the universe can be dated by examining carefully the scientific evidence, of which there is no shortage.

AGE OF THE EARTH

The popular answer is to be found in any encyclopaedia: the Earth is about 4.6 billion years old (and the universe is between 10 and 20 billion years old). End of story – except to scientists, who ask awkward questions, like where do these figures come from? Are they meaningful, particularly since they seem to have been revised repeatedly throughout the history of human knowledge? If they have so often been wrong in the past, are they correct now? Show us the workings!

Happy to oblige. At first glance the age of the Earth is the age of its oldest rocks, but things are never quite so simple. Superficially the Earth is hardly likely to be younger than its oldest rocks – but is that a hard-and-fast argument? A house is younger than its bricks, particularly if it is built of recycled bricks. It is not beyond all realms of possibility that a relatively young Earth has been peppered with large quantities of older rock during its existence. The surface of the Moon or Mars provides an impressive reminder that planetary bodies have suffered continual bombardment. As William Anders reported back from Apollo 8, the first manned spacecraft to fly close to the Moon in 1968, 'You can see the Moon has been bombarded through the aeons with numerous meteorites. Every square inch is pock-marked.' Calculations show that because of continual obliteration by fresh micrometeorites, lunar microcraters are never more than about one million years old.

The rocks now used to date the Earth's surface are assumed to have been melted and cooled, sorted and condensed on the Earth itself. Clearly, the Earth had to be in existence in some form before this recycling process began and in that sense the Earth is older than its oldest rocks. The solidification period was the time when the radio-isotope clocks of the rocks were reset, but the true origin of the Earth must have been earlier. How much earlier is open to question.

Before Charles Darwin's work on evolution, virtually everyone in western society took the date of the creation of the Earth from a literal interpretation of the scriptures. The oft-quoted timing of the Creation to 9 o'clock on the morning of 23 October 4004 BC ('just as the leaves were turning') is usually credited to (or blamed on) James Ussher's pronouncement made in 1650. It's easy to smile in the twenty-first century, yet Ussher was a learned man by any standards, professor of divinity at Trinity College, Dublin, later its vice-chancellor, archbishop of Armagh, primate of Ireland and accomplished author, among many other achievements.

Ussher was catapulted into posterity, although he did not live to know it, by the accident of having his calculated date of the Creation incorporated as a commentary note in the published Bible many years later. The story has improved a little with the telling and he in fact said the evening of 22 October, not the morning of 23 October, which is not going to have a devastating effect on the rest of this book. The refinement of Ussher's timing to 9 o'clock the next morning seems to have been grafted on later from a slightly earlier attempt (but younger Creation date) attributed to Dr John Lightfoot, another learned man, Cambridge University scholar of ancient languages and also later vice-chancellor. Indeed, numerous scholars over the centuries had attempted to calculate the date of the Creation with remarkably consistent – or fashionable – results of around 7000 years ago, give or take 1000 years or so.

In essence, the method was to work carefully through the Old Testament records of relationships all the way back to Adam and Eve, no doubt with a few indulgences about missing persons, average generation times and some extraordinary claims to longevity. The whole cake was iced with arbitrary constraints and special astronomical or astrological pleadings to make the Creation coincide with some significant celestial event when the dials were set to zero, as it were. People of different faiths understandably arrived at different conclusions. For the interested reader, early chronology is well reviewed by Brent Dalrymple (see Appendix).

By setting up a hypothesis that sooner or later would be questioned and indeed disproved, Ussher, Lightfoot and their predecessors, whether they realised it or not, were doing a splendid service to science. However, the way to a better answer is to find a better method.

Pick up a half-consumed cup of coffee and it will be somewhat cool, but pour some fresh from the same pot and it will still be hot. Why is the coffee in the pot hotter than the coffee in the cup when it was all brewed at the same time? The reason is that the larger volume in the pot has a smaller ratio of surface to volume and therefore loses heat more slowly. Looked at another way, in a large volume, the heat trying to escape from the centre warms the liquid further out, but from a small cup the heat is lost into space.

The eighteenth-century French Count Georges Louis Leclerc, Comte de Buffon, tried to apply the same simple principle to estimate the age of the Earth. He roasted a number of different-sized metal spheres to a high temperature and measured their rates of cooling. All he had to do then was to extrapolate the cooling rate from his small spheres to one the size of the Earth. Taking account of the Earth's composition, starting temperature and present temperature would reveal how long the Earth had been cooling and presumably how long it had existed. The idea was brilliant, but since nobody knew the starting temperature, present temperature or composition of the Earth, the escapade was mildly heroic and his result of 74 832 years would now be regarded as about 60 000 *times* too short (he also reckoned that life would end at year 168 123, so press on with writing your will).

Nevertheless, compared with archbishop Ussher's 4004 BC, it was a step in the right direction with the added respectability of an experimental foundation.

A century later the renowned physicist Sir William Thomson (who in that wonderfully British tradition metamorphosed into Lord Kelvin) based his calculations on the same principle but different data. The idea was still that the Earth had once been molten, so how long had it taken to cool to its present state? The observation that temperature in underground mines increased with depth hinted that the Earth was hotter towards the centre. This gave a clue to the rate of increase of temperature with depth, but any attempt to extrapolate 6000 kilometres or so through unknown materials all the way to the Earth's core from a mere kilometre or two of mine depth was bound to give a wildly approximate result. Taking into account the token melting point of rock and such measurements as were available of the heat gradient near the Earth's surface, but with most of the other unknowns supplied by guesswork, Kelvin came up with answers of around 100 million years, with a range from 20 million to 400 million to be on the safe side. As the suspicion slowly grew that even this upper figure was too short, later investigators were duly able to raise it, incidentally highlighting the weakness of a calculation fraught with so many imponderables, namely that the starting assumptions can acquire respectability because they support a particular result.

The age of the Sun was a companion interest to the age of the Earth, because it seemed likely that the solar system was formed more or less all at the same time, and in any case both the heating of the Earth and the existence of life needed the Sun (as far as people knew). Since the thermonuclear source of the Sun's energy was unsuspected at that time, the origin of its energy was assumed to be mechanical and the Sun was calculated in those days to be much younger than it is now considered to be. The calculation was based on the fairly simple principle that when objects collide through gravitational attraction, heat is released according to the mechanical equivalent of heat. This is the principle that energy cannot be destroyed, only converted into another form, so for instance a bicycle pump gets hot when used, as does a drill bit used to make a hole or two sticks rubbed together. Assuming on this basis that the Sun's energy had come from the collision of sufficient meteors to account for its mass, its energy content could be estimated. Kelvin's first estimate of the Sun's age by this method was just a few tens of thousands of years, but later he increased it considerably to something in the region of the magic 100 million years.

Yet another nineteenth-century method capable of giving an age of the Earth at around 100 million years was the science of stratigraphy, the measurement of rates of erosion and sedimentation. John Phillips in the 1860s used data from various areas of the world displaying major erosion and sedimentation, such as the River Ganges, to arrive at an age of a little less than – guess what – 100 million years. Once again, any such calculation is based on a large number of unknowns and in any case dates back only to a starting point for a particular geological feature on a pre-existing Earth. The underlying planet has to be older.

Darwin himself used the sedimentation and erosion method, albeit with rather sparse data, to arrive at an estimate of about 100 million to 300 million years. His back-of-an-envelope figuring for the *Origin of Species* (1859) is a magnificent example of the incredible power of common-sense approximation (the calculation is very easy and is reconstructed in the Appendix). Darwin simply toyed around with the probable rate of erosion of chalk cliffs by the sea and settled on a working figure of an inch (about 25 millimetres) of erosion per century for cliffs 500 feet high (about 150 metres), taking into account that they collapse through undercutting from time to time. He then reckoned that the Weald of Kent and Sussex, carved out between the North and South Downs not far from London, represented 22 miles (about 35 kilometres) of similar erosion of cliffs that were originally 1100 feet high, arriving by simple proportion at an age of 306 662 400 years. Finally, he graciously allowed that such an estimate was bound to be just that, an estimate, and was ready to accept a lower and rounder figure of around 100 million or 150 million years, or on the other hand a 'not improbable' period of more than 300 million years.

The Mississippi delta, the Columbia river, the Channel between England and France and many other geological formations have been studied in a similar way, but the weakness has invariably been the lack of an adequate measurement of either erosion or sedimentation rates.

Approximations in science are nevertheless very valuable. It doesn't matter that Darwin's estimate allowed at least a threefold latitude from 100 million to 300 million years or more, or that his contemporaries treated the calculation with derision. Nor does it matter that the age of the Earth is now reckoned to be about 15 times greater than Darwin's higher estimate from the erosion of the Weald. What does matter is that the calculation was far nearer correct than 4004 BC, or 74 832 years, and was based on such observations as were available. Darwin's common-sense result spanned the now accepted answer of about 135 million years for the Weald and he clearly acknowledged that this localised formation hardly represented the entire history of the Earth.

The increasingly satisfying consensus age of around 100 million years or so in the late nineteenth century received further boosts to its credibility. T. Mellard Reade had the idea of measuring the chloride and sulphate concentrations in sea water, and John Joly did the same for sodium. The reasonable assumption behind these ideas was that the oceans came into existence when the Earth was sufficiently cool for water vapour to condense and fall as rain. Sea salt had built up ever since through the cycle of water washing salt off the land and then evaporating from the oceans as pure water, leaving behind the salt. By measuring sea and river salt concentrations, the time taken for this to happen could be estimated. Once again, the calculations came out broadly in support of a 100-million-year timescale. To be fair, Joly did express doubts about the reliability of his own findings.

Nothing is more comforting in science than to get the same result from several different methods. The convergence on an age of around 100 million years from considerations of sedimentation and erosion, the cooling rate of the Earth and the salinity of the oceans must have been very convincing to late-nineteenth-century scientists. It was also very beguiling. Books on the subject written a mere century ago are truly fascinating for two things. One is the cosy sense of consensus as people plugged different numbers into equations for the age of the Earth, inevitably finding good reasons to prefer those numbers that gave ages of around 100 million years. The other is the tone of ultimate rightness – not to say righteousness – applying to the choice of those numbers, an attitude that the reckoning had been all wrong in the past, but with the benefit of then-modern philosophy the author's fresh opinion was beyond risk of contradiction. Here is Samuel Laing in full flood in 1890, in *Problems of the Future and Essays*, one of those wonderful theory-of-everything works ranging over atomic theory, anthropology, poetry and taxation (to mention but a few): 'Lyell, and a majority of the best geologists, consider that one hundred to two hundred million years are required to account for the *undoubted facts* of geology since life began' (my italics; below also).

Laing, following Kelvin, led readers through the supposedly indisputable history and future of the Sun in some detail. Don't get hung up on the figures, they are quoted only to show how completely wrong his principles were. First he showed how a square metre of the Sun, and the matter beneath it down to the centre of the Sun, would weigh 224 000 000 tons. Then he used the principle mentioned above, that matter crushing in towards the centre of the Sun would generate heat, and took into account the heat output from the Sun, which scientists of the time could measure. Laing then said: 'But the radiation from each square metre of the solar surface in heat per hour is equivalent to 78,000 horse power ... An *easy calculation* shows that to supply energy at this rate for a year, our supposed cone of 224,000,000 tons must fall one metre in 313 hours, or about 35 metres a year.' Horses, one surmises, were dragging their feet over metrication – however, it followed convincingly that '...if the Sun's radiation of heat has been uniform for the last fifteen millions of years, the solar radius must then have been four times greater than it is now; and if the present supply were maintained by shrinkage alone, for the next twenty millions of years, the Sun must have shrunk to half its present size.'

It doesn't matter how purportedly *easy* the calculation is, or how nearly unanimous the consensus, if the starting point is hopelessly wrong. Which it was, because the Sun does not generate heat mechanically through 'shrinkage alone'. Around the corner was a prime example of T. H. Huxley's 'great tragedy of Science – the slaying of a beautiful hypothesis by an ugly fact' (*Biogenesis and Abiogenesis*).

RADIOACTIVITY DISCOVERED: START AGAIN!

The Curies entered the arena and slayed everything by discovering the ugly fact of radioactivity. All calculations based on the rate of cooling of the Earth were suddenly rendered meaningless if heat was being generated by a previously unsuspected internal source, which was radioactive decay. It's no use measuring the cooling rate of a planet that may not be cooling. Ernest Rutherford (Lord Rutherford of Nelson) soon calculated that the heat being replenished by radioactive decay would cause the Earth to remain hot for much longer and therefore it could be much older than the sacrosanct 100 million years.

Radioactivity has played an enormous part in unravelling the ancient history of life, the planets and the universe, partly by accounting for a new source of heat, partly by providing an exquisite dating system, and partly by leaving traces of the way life has treated different atoms. It is often the case in science that the true discoverer of something new is difficult to pin down because each discovery may have been influenced by an earlier observation, the significance of which was not fully appreciated at the time. Credit tends to go to the person who realised the full significance and developed the study. So it was with radioactivity and the Curies.

What happened was that in 1895, Wilhelm Roentgen discovered X-rays, as he called them to signify their mysterious nature, and for this he later received the Nobel Prize. X-rays caused certain chemicals to glow (fluoresce) and fogged a photographic plate as though it had been exposed to light (it was black when developed). This was a momentous discovery, because it was quickly realised that bones and flesh placed in the X-ray beam stopped different amounts of radiation, creating a picture of the inside of the living body. Almost immediately after Roentgen's discovery of X-rays, Antoine Henri Becquerel discovered that mysterious rays emitted by the element uranium could do the same thing. The discovery was accidental. Becquerel was experimenting with the observation that crystals of a uranium compound, after exposure to sunlight, fluoresced and would fog a photographic plate. One day he took a wrapped photographic plate out of a drawer where it had been left near some of the crystals that had not been exposed to sunlight – and on development it was already fogged. Uranium was clearly emitting some sort of rays that, like X-rays, could penetrate the wrapping and fog the film.

This observation was in turn followed up by Marie Curie and her husband Pierre Curie (whose brother Jacques had previously helped design an electrometer essential to their work). Nothing they could possibly do to uranium, no chemical or physical effect such as temperature, changed the radiation coming from it in any way, therefore radioactivity was a property of the uranium atoms themselves. Radiation could be partially blocked with materials such as lead or bones, but the emission from the uranium itself was unaffected. This discovery has turned out to be incredibly important, because whatever conditions radioactive elements have been subjected to in the history of the Earth and even in remote galaxies, where they can be detected, nothing has changed them or the rates at which they are transformed from one element into another (except induced nuclear reactions), giving away their ages.

Becquerel and the two Curies were jointly honoured with a Nobel Prize, and by having physical units named after them. For many years the unit of radioactivity, as disintegrations per second (clicks on the Geiger counter, if it could catch them all), was the Curie, which was based on a gram of radium 226, the highly radioactive element also discovered by Marie Curie. The standard Curie was 37 billion disintegrations per second, a colossal amount of radioactivity in laboratory terms. The more modern Système Internationale (SI) unit has been set at one disintegration per second and named after Becquerel.

The newly discovered source of heat from radioactivity did not complicate the calculation of the Earth's age but in fact simplified it. Rough calculations of cooling were no longer necessary, because radioactivity actually provided a completely independent and much more accurate method of dating. Radioactive elements decay to become lighter elements, and do so at a fixed rate. For instance, radium 226 decays to the gas radon 222, emitting alpha particles and gamma rays, at a rate that halves the remaining amount of radium every 1620 years (the 'half-life'). Early in the twentieth century the sequences of decay from element to element, and the half-lives of the radioactive elements, began to fall into place. Measuring the half-life doesn't mean standing around waiting for it to happen, of course; counting the radiations from a known weight, and therefore number, of atoms allows it to be calculated.

Working backwards from the radioactive elements and decay products found in a piece of rock (by measuring their masses), it is possible to calculate how long the process has been happening and hence the age of the rock. One of the best elements for dating the Earth is uranium, because traces occur quite widely and one of its isotopes, uranium 238, has a very long half-life of almost 4.5 billion years. The principle is that if rock were originally formed containing 'fresh' uranium 238, which decays eventually to lead 206, then the ratio of these two elements present today allows year zero to be calculated. Needless to say, the actual science of radiometric dating is vastly more sophisticated than it sounds from this simplistic description. Sources of error are understood: for example, if some decay product had been there at the start, the estimate of age would be too high and the rock could really be younger. Another complication is that although nothing can affect the rate of radioactive decay, the mineral products themselves can be moved around and resorted by the effects of heat. Far from this leading to an error, the interpretation of such movements through the science of thermochronometry can give positive information about the history of a rock.

Where possible, alternative isotopes are analysed as an independent check. Rubidium 87, which decays to strontium 87 over a half-life of 48.8 billion years, is widely used and tends to agree with uranium/lead dating. Much longer-lived isotopes are known, such as samarium 147 which decays to neodymium 143 with a half-life of 106 billion years, but they are rare elements and are not so useful when the amounts present are too small for accurate measurement. At the other end of the scale, if an element has a short half-life, all of the original isotope will have decayed by now and the ratio cannot be measured.

Large numbers of rock specimens from all over the world have been tested by uranium/lead or rubidium/strontium dating. Many rocks have ages greater than 2.5 billion years and the oldest date to about 3.9 billion years. Very recent technology with lutetium and hafnium ratios suggests an antiquity nearer to 4.1 billion years.

In one sense, whatever the age of the Earth, its material is all the same age, which is the age of the universe. This has, of course, been considered. When rocks are formed the chemical components are effectively resorted by processes such as crystallisation and in so doing the isotopic ratio clock is reset. The measured age gives the date when that particular rock was formed, not the date when the Earth was formed, which had to be earlier.

So, dating a piece of rock to around 4 billion years does not completely solve the question of the age of the Earth, but rather sets a minimum age and reformulates the question as: 'How did certain rocks apparently form, 4 billion years ago?' The answer depends on how the Earth was formed. Ancient rocks such as granite show every sign of having solidified from a molten state. How the material came to be molten is a different question; the key point is that the inclusion of metals and minerals in the fresh rock involved natural processes of separation, such as distillation or crystallisation, so that elements in the molten mixture were deposited relatively pure or enriched. Prospecting for minerals is all about finding these rich deposits.

AGE OF THE SOLAR SYSTEM

Logically the Earth is older that its oldest-dated rocks, but by how much? If uranium was being resorted and deposited in ores 4 billion years ago, there must have been an earlier stage. A favoured theory is that our Sun's planets accreted from dust and debris that collapsed through gravitation, becoming molten owing to the heat released by compression, friction and radioactivity. The evidence does not suggest that the Earth is much younger than the rest of the solar system. Nor does it suggest that the Earth is older, which could have meant life having more time to develop before the planet was somehow captured by the Sun.

Extraterrestrial materials such as meteorites (meteors that have landed) can be picked up and dated and many are about the same age as the Earth. A number have been dated to about 4.5–4.6 billion years by the argon, rubidium, potassium and lead isotopes that they contain. Rocks brought back from the Moon have also proved to be in the 4 billion year bracket. However, extraterrestrial rock samples may give a misleadingly young age for themselves because they must have been through some previous existence in which atoms and dust formed into chunks of rock, by gravitational accretion, melting or both. They must have existed in the form of a sizeable asteroid or planetary body that subsequently broke apart, because a small cloud of dust with little gravitational interaction would not accrete into hard rock. Some meteorites contain high concentrations of metals in a form that can only indicate a previously molten state and therefore a substantial size. Recent discoveries suggest that the decay of radioactive isotope aluminium 26 could have been a major source of heat. The proviso about the age of Earth rocks being a minimum applies equally to meteorites: isotope measurements indicate the time since the rock solidified, but give little indication of how long its mother material may have been in existence elsewhere in the universe.

If the Earth accreted from interplanetary dust and debris until it was large enough to become molten, the surface might have changed from solid to molten as the planet heated and initial bombardment continued, then back to solid as it cooled. Realistically, since life is water based, it is unlikely to have made significant progress before the surface and the atmosphere had cooled to the temperature of liquid water.

On the other hand, it is possible that there always was a solid surface. If the Earth formed from debris and meteoritic material, the incoming material would initially have been extremely cold indeed, little above absolute zero, because being much smaller than planet sized the pieces of debris would not have held much heat and space is extremely cold. The planet had to heat up from cold, by the release of the mechanical equivalent of heat following gravitational impact and from radioactive decay, but it had to be planet sized to do so. The arithmetic of surface to volume ratios, mentioned a few pages earlier in the analogy with coffee pots and cups, means that a large body such as a planet has proportionately little surface area and loses its inner heat slowly. The temperature would always have been lowest at the crust where heat radiated to the fierce cold of space. The Earth at that time is believed to have been under continual bombardment from interplanetary debris that caused cataclysmic disruption and surface heating, but the word 'continual' needs to be interpreted in its astronomical sense; the majority of the Earth's surface at any particular time may have been relatively unscathed. Much of the surface need never have melted despite the gradual formation of a molten core. This would especially have been so before the growing Earth had gathered sufficient mass to retain a dense, insulating atmosphere. It would also help to explain why the Earth never lost all its water.

Zircons (zirconium-containing mineral crystals) found recently in Australia have been dated to 4.2–4.4 billion years old by isotope measurements. These samples are quite special because, according to their oxygen isotope content and some intricate reasoning, it is possible to deduce that they had melted and interacted with rock that had itself reacted with liquid water, meaning specifically water at low temperature rather than water vapour or steam. In other words, some liquid water was present on the Earth as long as 4.4 billion years ago. If water could survive, so might life.

The solid surface of the early Earth may have been extremely thin and indeed it still is. We happily survive nowadays floating on the top 1 per cent of a globe of white-hot magma, as is instantly obvious when it gushes out somewhere. There may well have been a period when the surface temperature peaked, the crust was flimsier and cracking was more prevalent, but if the planet began as an accretion of dust and debris, then conceivably it always did retain a solid surface. The minerals now excavated for dating may be more recent outpourings onto an older surface, particularly since uranium and lead are among the densest elements and would gravitate naturally downwards rather than upwards.

The match of the age of the Earth to the solar system rules out various distractions such as life on Earth having spread from older local planets (or moons) where it might have had more time to develop. The shortage of time on the early Earth cannot be bypassed by proposing an older solar system. Can the universe help?

AGE OF THE UNIVERSE

The age of the universe (between 10 and 20 billion years) has been deduced quite differently from the ages of the Earth and the solar system. The basis of the calculation is that the universe is expanding in all directions as though it originated in an explosion at a single point in space and time, the Big Bang as Fred Hoyle called it. If the rate of expansion has always been the same, it is only necessary to measure it now and reverse the expansion, whereupon everything in space should return to a single point on a single date. There are some philosophical flaws in this argument. It's no use looking for the place in the universe where it all started, because that point has expanded to the dimensions of the universe: it is everywhere. The Big Bang did not so much happen in space, but created space. Furthermore, extrapolation back to a certainty, a single point in space and time, might contravene Heisenberg's Uncertainty Principle, which says that one could narrow down only a probable range of location and time. The Big Bang might have been slightly damp and more like a Big Fizz, and the extrapolation idea should be taken no further than that the primordial universe was very small. Fingerprints of the nascent universe remain, billions of years later, such as the rotation preserved in distant galaxies indicating rapid rotation of the early universe.

Fred Hoyle's alternative theory, the Steady State, suggested that the universe is in a state of equilibrium whereby fresh matter is continually coming into existence but is balanced by other matter that is continually disappearing. Steady State is out of favour and Big Bang is now widely accepted, although it is worth remembering that the correctness or otherwise of a theory is not influenced by a show of hands. It is also a good scientific principle not to discard a theory because there is good evidence for an alternative. Both can be right. A Steady State universe could have been the origin of the material that collapsed gravitationally to precipitate the Big Bang. Some sophisticated models of the so-called vacuum of space suggest that subatomic particles are continually appearing and disappearing, a form of steady state.

The universe is vast and contains unimaginable quantities of material, so how and when can it originally have been very small, as cosmologists say? 'How?' is a difficult question, but our inability to explain how something happened is not evidence against it happening. The more answerable question is: 'When?' The essence of the date calculation is disarmingly simple and is based on changes to the wavelengths of light emitted by stars. Here's an outline of the argument.

SEEING THE LIGHT

Most of what we know is what we see. And seeing is the result of light interacting with our eyes. Numerous technologies have been invented to enhance and analyse light before it reaches our eyes, particularly microscopes, telescopes and spectroscopes. The last of these is a device that sorts out the different colours and signals mixed up in a light beam. Let's have a brief look at what light really is.

Light is electromagnetic radiation, or energy, which occupies a continuous spectrum of wavelengths (or frequencies of vibration, which are related inversely to wavelength). The human eye can detect electromagnetic radiation having wavelengths between about 400 and 700 nanometres (nm), which is the sole reason why this band of wavelengths is treated as special and given the name 'light'. Sunlight is really a mixture of the colours of the rainbow ranging from violet (around 400 nm) through blue, green, yellow and orange to red (around 700 nm). Not all colours are present at equal intensity, but the particular mix to which the brain is accustomed is thought of as white.

Daylight can vary considerably, for instance becoming more yellow or red when the Sun is low in the sky and its light passes through a greater length of atmosphere. The brain can adapt to a colour cast in a matter of minutes, even quite a strong colour such as the effect of tinted sunglasses, and if we had grown up with a different mix of daylight wavelengths we would still regard it as white. A simple indication of this is to be found in colour-blind people who lack the ability to detect certain colours, depending on their type of colour blindness. They still interpret their limited mix of daylight colours as white.

A rainbow is the effect of the wavelengths, or colours, contained in sunlight being separated out as a display, known as a spectrum. All the colours of the rainbow are present in sunlight in the first place. A spectral display is generated because as light passes from one substance to another, as from air to raindrops, the different wavelengths of light bend slightly differently and are separated. The same effect is achieved when light passes from air to glass and can be noted around the house when light strikes the prism-like edges of a cut-glass vase or the bevelled edge of a mirror. The scientific instrument known as a spectroscope (spectrometer, spectrograph) allows the separated colours to be inspected very closely and is calibrated so that the wavelength of any colour can be measured in nanometres.

It has been known for two centuries that when sunlight is analysed in detail with a spectroscope, the spectrum does not appear exactly continuous from violet to red as might be expected, but is interspersed with a large number of fine black lines, or breaks, irregularly spaced. The lines are named after Joseph von Fraunhofer. He didn't actually discover them – William Wollaston did in 1802 – but Fraunhofer did a more detailed study and arrived at a better explanation for the lines in 1817.

The reason for these dark lines in the solar spectrum is that chemical elements can have colours. Immediately one can devise a way of detecting chemicals in the Sun and stars. When atoms are excited by injection into a high-temperature flame or by an electric discharge, internal rearrangements cause them to emit light. Fluorescent lighting tubes emit sharp bands of green, yellow and violet. A piece of copper wire thrown on a fire causes the flames to be coloured green. Driftwood from the seashore burns with a rich yellow flame owing to the sodium in sea salt, the same colour as sodium street lamps. Potassium burns with a lilac colour, strontium burns bright red, to the delight of fireworks enthusiasts.

However, these colours are not just vaguely yellow or green or red, they are extremely sharp spectral bands of fixed and discrete wavelength caused by the emission of exact quanta of energy when electronic transitions occur inside the atoms. The wavelengths are precisely known and are freely published. Because certain elements emit characteristic wavelengths, or colours, conversely the detection of those bands in a spectrum shows that element to be present. If the red wavelength special to strontium is detected, then strontium is there.

Back to the dark lines in the solar spectrum. These have wavelengths characteristic of iron, calcium, sodium and many other familiar elements. The reason the lines are dark rather than bright is easily demonstrated by an experiment back on Earth. A Bunsen burner gas flame is lit, and if the air flow is adjusted correctly the flame is an almost colourless pale blue. The gas flame is then tinged with colour by burning, say, sodium salt, which would make it bright yellow. A spectroscope is set up to look at the flame and the well-known sharp yellow lines are observed. Next, the spectroscope is turned away from the flame and towards an incandescent white light bulb, whereupon a continuous spectrum from red to violet is seen. Finally comes the exciting part, which is to combine the two observations. The sodium flame is placed between the white light and the spectroscope. Dark lines now appear in the continuous spectrum, exactly where the sodium lines previously were. The reason is that the reaction causing sodium to emit yellow light is also able to absorb at exactly the same wavelengths, and it absorbs yellow bands from the continuous spectrum, leaving dark gaps. In fact the yellow sodium lines are still there but are too dim to see.

That is what happens in the Sun. The intensely hot interior is equivalent to the light bulb. Here vast numbers of emission lines are generated, but are broadened and blurred into a continuous spectrum because the particles of plasma are moving about at great speeds, causing a Doppler effect (described shortly). The outermost regions of the Sun, what might be called its atmosphere, represent the Bunsen burner. Here the gases are much cooler and elements exist in the form of vapour, like the sodium in the gas flame. In this condition they emit their tell-tale wavelengths but also absorb them, and since the source of the continuous spectrum in the Sun is much brighter than the emission lines, the dark absorption lines appear.

Most importantly, the characteristic lines prove that the Sun, and also vastly distant stars, contain exactly the same kinds of elements as are found on Earth. In addition they reveal that the mix of chemical elements is different in different types of stars, giving clues to their chemistry, temperature and history.

Two important scientific tenets are, first, that the wavelengths of the lines from a particular type of atom, at the time they are generated, are the same anywhere in the universe; and second, that those wavelengths have always been the same throughout time. If either of those assumptions should turn out to be wrong, cosmology is in deep trouble.

Now let's see how all this leads up to the age of the universe.

In light from distant stars the lines are dutifully present, except that their wavelengths are longer than they ought to be. Having just said that the wavelengths *when generated* have always been the same anywhere in the universe, somehow they must have changed. The familiar clustering of key lines is easily recognisable in the starlight and there is no reason to doubt that the lines were in the right places when they were generated on the star, but they have all moved towards the red end of the spectrum, like a musical chord played in a lower key. Crucially, the more distant the star, the further the shift of the lines towards the red.

This phenomenon is known as red shift and its importance was first appreciated in 1929 by Edwin Hubble, in whose honour the Hubble orbiting telescope was later named (as so often in science, the effect may have been discovered earlier, in this case by Vesto Slipher in 1920, but Hubble developed the significance of the discovery).

The accepted explanation for red shift is that the star is travelling away from us at high velocity. There is a well-known analogy with sound. The sound of something approaching reasonably fast, such as an ambulance siren, train whistle or low-flying aircraft, drops in pitch (as distinct from loudness) as it passes by and recedes into the distance. This effect is named after the person who first investigated it, Christian Doppler. It is important to note that the speed of sound through air is constant, under particular atmospheric conditions; sound from a receding source does not travel any slower, but it sounds different.

Something similar happens with light. As Albert Einstein pointed out, the speed of light in a vacuum is constant anywhere in the universe. But its pitch or frequency drops (wavelength increases) if the source of the light is moving away from the observer (or space itself is expanding); conversely the frequency rises if the source is moving towards the observer. For a colour somewhere between red and blue, this means a shift towards red if the source is receding or towards blue if the source is approaching. The greater the speed, the greater the shift of wavelength, giving a convenient means of measuring the velocity of the source relative to the observer.

Hubble, observing spiral nebulae, realised that the more distant the star, the greater its red shift and the greater the apparent velocity with which we and the star are moving apart. It doesn't matter in which direction one looks in the sky, stars can be seen, and the further away they are the faster they are receding (in a general sense; a galactic arm, for instance, might be rotating towards us). This doesn't mean that the Earth is stationary or at the centre of the universe and that stars are receding specifically from the Earth, it means that the stars and galaxies of the universe are moving away from each other at colossal speed. The idea developed that the universe is expanding in all directions as though it originated at one point, but as mentioned earlier, the Earth is not that point and neither is anywhere else. Reverse this motion, extrapolate back to a point, and what happens? A big splunch coinciding with the moment of the Big Bang, the origin of the universe.

Other explanations for red shift have been suggested. Franz Zwicky had the early idea of light from distant stars becoming 'tired' through losing minute amounts of energy after travelling for billions of years, but that is no longer supported. The same applies to suggestions that spectral lines could have been created redder in the distant past, or that red shift is caused by natural laser action. Einstein did, however, predict a red shift, since verified, in an intense gravitational field such as is generated by a white dwarf star (or indeed in any gravitational field, to a miniscule extent).

To track down the date of the Big Bang another essential piece of information is needed, which is the precise distance of the red-shifted stars. Actually this is not their present position but where they were millions of years ago when they sent out the light that is now reaching Earth. In a sense we can see into the past, which seems rather clever except that the penalty is that we can see *only* the past and have no idea what the universe is doing at the present (how the universe may have changed since light left a star billions of years ago is a quagmire of imponderables best not ventured into). This is not something peculiar to star light; if you look at someone a metre away, you see them as they were in the past, one 300 000 000th of a second ago (and hear them as they were one 300th of a second ago).

Unfortunately, it is difficult to be precise about the distances of most stars or galaxies. The calibration scale is based on an edifice of measurements, known as the distance ladder, in which distances of relatively close stars are used to estimate those of more distant stars and so on, extending eventually to the fringes of the visible universe. Many researchers contributed to the establishment of this scale, notably Henrietta Swan-Leavitt working on stars of variable brightness; Harlow Shapley, who applied these considerations to the dimensions of our galaxy; and Edwin Hubble, mentioned previously.

Only for the closest stars is there a prospect of measuring distance by extremely accurate surveying principles (triangulation). Even then, the survey baseline has to be the scale of the Earth's orbit around the Sun, by measuring the minute shift in the apparent direction of the star over a period of months. Most stars are so far away that this shift is too small to measure. The star appears never to move, not even a minute amount.

Distance thus has to be estimated indirectly. One method is to deduce the star type from its chemistry, revealed by analysis of its light spectrum as mentioned previously, which in turn tells us what brightness to expect, on average. Comparison of the apparent brightness of a star with similar-type stars at distances close enough to measure gives a guide to the unknown distance, provided that there is not a cloud of dust in the way to make the star appear dimmer (but that can be detected because the dust absorbs some wavelengths more than others).

Another method is to assume that galaxies of a particular type are all about the same size, on average. If the probable size of a galaxy can be estimated from its type, and its image in the telescope can be measured, then the smaller the image, the further away the galaxy is, and this distance can be estimated. A measurement of the red shift of stars in the same galaxy gives their relative velocity.

Either type of distance estimate – star brightness or galaxy size – is obviously open to substantial error and involves a certain amount of circular argument. If distance is calculated inaccurately then a similar inaccuracy strikes at calculations of the expansion rate of the universe.

A combination of star velocity (deduced from red shift) with star distance (deduced from brightness or galaxy size) allows the expansion of the universe to be calculated as one single convenient number, known as Hubble's constant (H_0, the $_0$ signifying the present-day value). The consensus Hubble constant is about 70 to 80 kilometres per second per megaparsec of distance, which is vastly simpler than it sounds. All it means is that a star receding from us at 75 kilometres per second is about a megaparsec distant; one receding at 150 kilometres per second is 2 megaparsecs distant, and so on. It's easier still to understand because a megaparsec is simply a convenient measure of distance, used to avoid big numbers. Interstellar distances are so colossal that kilometres soon overflow with zeros, so parsecs are used instead.

The word parsec comes from parallax-second. As mentioned earlier, a close star appears to shift by a tiny angle, compared with 'fixed' stars at 'infinite' distance, as the Earth moves in orbit around the Sun. By simple geometry, a star that appears to shift 1 second of arc (1/3600 of a degree) over a baseline corrected to 1 astronomical unit (the Earth's distance from the Sun) is 30 900 billion kilometres away, and this distance is known as a parsec. Dividing by the distance light travels in a year, a parsec is 3.26 light years. Even a parsec is too tiny for many astronomical distances, so astronomers call a million parsecs a megaparsec, which equals 3 260 000 light years, or about 30 900 000 000 000 000 000 kilometres (you see why astronomers prefer to say one megaparsec).

In theory, Hubble's constant leads to a straightforward date for the Big Bang. The calculation is breathtakingly simple (see Appendix, which also explains a short cut). Although Hubble's constant (H_0) is usually given in kilometres per second (recession) per megaparsec (distance), this can easily be changed into *kilometres* per second *per kilometre* of distance, whereupon the '*kilometres*' and '*per kilometre*' cancel each other out. H_0 is left in simple units of 'per second'. Invert H_0 so that it becomes $1/H_0$ and 'per second' becomes inverted to 'seconds per'. Seconds per what? Seconds per the universe! In other words, the age of the universe in seconds. For $H_0 = 75$ the answer works out to be 0.4 million million million seconds, and since there are about 30 million seconds in a year, that's 13 billion years. Nothing could be simpler. The universe is 13 billion years old, if the value of 75 for Hubble's constant is correct.

Since relative recession velocity increases with distance from us, there is a distance at which it would equal the speed of light. That distance is 13 billion light years if H_0 is 75 (see Appendix). Accepting that an object cannot be observed travelling faster than light, we cannot see anything further away than about 13 billion light years (4000 megaparsecs) or whose light has taken longer to reach us than 13 billion years.

In practice, Hubble's constant is no more certain than the star distance estimates used to calculate it. Distances estimated from star brightness or galaxy size might be a little woolly, although the other ingredient, red shift, is certainly accurate. Measurements of Hubble's constant ranging from 50 to 100 have been considered respectable, giving ages for the universe ranging from 20 billion years down to 10 billion years. Contemporary measurements converge on about 70–80 for H_0 and of course every new measurement tends to come with an implication that it is the last word. In recent months an age of 13.7 billion years has gained support.

If all this sounds too simple to be true, you might be right. The simple inversion of Hubble's constant to change H_0 (per time) into $1/H_0$ (time) would be correct if the universe has expanded at a constant rate. Provided that the universe contains less than a certain amount of matter (or energy), it may well have expanded at a constant rate. Such a universe is called 'open' and its age is simply $1/H_0$. But if the universe contains more than that certain amount of matter, gravity would be sufficient to counteract its expansion, causing it to slow down. Such a universe would ultimately stop expanding and begin to collapse back on itself, so is called 'closed'. Since in the past it was expanding faster, it took less time to reach its present state and is therefore younger than calculated for an open universe. How much younger depends on the precise model, but a favoured figure is two-thirds as old. The formula for its age would be not $1/H_0$ but $2/(3H_0)$.

Slipped in between the open and closed universes is a model called the 'flat' universe, where the one model borders the other. In the flat model the amount of matter is the maximum, and the rate of expansion the minimum, that together will exactly prevent gravitational collapse. It amounts to having it all ways: expansion is decelerating but will continue for an infinity of time. The model is close to an open universe and $1/H_0$ is near enough.

So which model of universe do we live in: open, closed or flat? The story keeps changing as modern technology turns up fresh clues. A comprehensive *Science* article on the question is listed in the Appendix. The closed model has considerable support, but one difficulty is that the amount of matter actually visible in the universe, according to astronomical observations, is insufficient to lead to gravitational collapse, therefore an open model ought to be favoured, unless perhaps the majority of matter (and energy) is so-called dark matter that cannot be seen. The flat model has an attractive elegance about it that implies a perfect balance between expansion and gravitation, but on the other hand it represents a special case.

Recent evidence could suggest a revolutionary new model in which the expansion of the universe has actually been accelerating since about 5 billion years ago. The evidence for this is that supernovae of a certain type at extreme distances have less red shift, and therefore are receding more slowly, than would be expected for their distances. Distance is estimated from brightness and is less reliable than red shift, but provided that the distance estimates are on average correct, the inevitable conclusion is that expansion was slower in the distant past when the light left these supernovae. A different method, analysing the scatter of matter and background radiation in the universe, also suggests a possible acceleration model.

If the accelerating expansion is real, its cause is a moot point. Mathematically the cause is a small constant in an equation, signifying repulsion of matter. The constant is similar to the well-known cosmological constant devised by Einstein in about 1917 to overcome gravitational contraction and so maintain a static universe. Einstein's constant seemed to become redundant when the universe was shown to be expanding. Einstein then regretted introducing the constant and used to refer to it as *mein grösster Fehler* (my greatest mistake). In its new form the constant explains acceleration in an already expanding universe. The aesthetic problem is that the need to invoke an arbitrary constant makes the theory more complicated, whereas the usual sign of scientific progress is a simplification.

There are implications for biology in a universe of accelerating expansion. If the expansion rate was once lower, it is now higher. Because an accelerating universe would have taken longer to reach its present state and is therefore older than $1/H_o$, life had more time to emerge.

All of the above arguments date the universe from the time it has been expanding. Recently a completely independent method, based on the actual age of an ancient star, has indicated a comfortingly similar age of around 12–13 billion years. As pointed out earlier, spectral lines in starlight are the unmistakable signatures of chemical elements present in a star's atmosphere. With the increases in sensitivity and resolving power made possible by recent technology, more elements are being detected in the atmospheres of stars. The signatures of some radioactive isotopes can be seen, such as uranium and thorium, which must have fixed decay rates and half-lives just as they do on Earth. Although the reasoning is intricate, one international team has used the method, termed radioactive cosmochronology, to set an age of around 12.5 billion years for a very old star indicative of the age of a galaxy and, by implication, a minimum age for the universe (a summary by Christopher Sneden is referenced in the Appendix).

The completely different methods for calculating the age of the universe and the age of the Earth reassuringly show the universe to be older than the Earth, thereby avoiding some embarrassment. But shouldn't they be the same? Isn't all the universe the same age, including the Earth? A parallel argument was raised earlier when saying that many Earth rocks are younger than the Earth, meaning that they were melted and reformed after the Earth came into existence. Similarly, the Earth is younger than the universe because the Earth is dated from the time it formed from recycled universe material, its elements were resorted and its isotopic decay clocks were reset.

PAINTING LIFE INTO THE UNIVERSE

Life took a long time to develop. If the question is where did it happen, the answer obviously has to be where there was enough time. Two pretty convincing times have emerged from the research of recent decades: around 12–13 billion years as the age of the universe, and about 4.6 billion years as the age of the Earth (with a suitable surface for the past 3.9 billion years). Each of these timeframes provides an empty canvas on which to try to paint the entire history of life. But which size best fits the picture?

The only way to approach any scientific problem is with an open mind, looking to satisfy the general case. This particular general case says quite simply that life can originate anywhere, anytime; indeed everywhere, all the time. The universe is the correct canvas and all the time in its history has been available. Even such an extreme environment as the atmosphere of a star can be seen as a factory for producing useful chemicals leading to life, while the intense cold of space near absolute ensures their preservation. No extreme, no process available in the universe can be precluded from playing a part in life's assembly process. Only after fitting the history of life to the universal canvas does it make sense to enquire whether it might have fitted within an Earth-sized frame.

The back-to-front way of going about it is to try to cram life on to an Earth-sized canvas just because it has an attractive frame. If it genuinely fits, all well and good, but we're looking for the full-length portrait, not cut off at the head and shoulders. The Earth is not necessarily the starting point for its own life because there simply isn't any reason why it has to be, other than special pleading to accommodate narrow preconceptions, and science abhors special pleading. Still narrower variants of this special case claim that life does not exist anywhere else in the universe because it never could, life is unique to this planet because the conditions on Earth are uniquely suitable for life, none of the billions of planets in the universe is remotely similar to Earth, the event that initiated life was unique, it was supernatural, and so on.

Dating life will be rather more difficult than dating the Earth and the universe, but nothing is impossible.

FRIENDS AND RELATIVES

All life on Earth is related to a single ancestor

The next little problem is to fit the history of life into the timescale of the Earth and the universe outlined in the previous chapter. If life on Earth took hold only once, the problem becomes much simpler. Only a single date needs to be found. Just as the expansion of the universe can be reversed to find the date of the Big Bang, so the evolution of life into the thousands of species alive today can be reversed to find the date of the first life. At least, that's the theory. It really has two components: the first key question is whether all life really is related to a single ancestor, which this chapter will address. And if so, the second question will be, how long ago did the single ancestor live?

APPEARANCES

What's the difference between an elephant and a goldfish? Superficial differences are more than obvious but as soon as we scratch beneath the surface the similarities begin to outweigh the differences. Bird, fish, elephant, all have a skeleton that is easily recognised as vertebrate, characterised by a vertebral column or backbone encasing a central nervous system connected to a head and brain. The nervous system connects the brain to the limbs to enable sensation and control. The front legs of the elephant or rabbit, the wings of the bird and the front (pectoral) fins of the fish are all anatomically related, being limbs attached to the equivalent of the shoulder (although fish fins are more for balance than propulsion). The leg of the bird, the hind legs of the quadruped or the pelvic fins of the fish are similarly attached to what humans would think of as the hip.

Humans walk on what they regard as the whole foot, from heel to toe. This human view of normality is not shared by many of our fellow vertebrates. Cats and dogs walk on their toes, the equivalent of their heel being the little lump found nearly halfway up the leg. The horse's hoof corresponds to even less of the human limb, merely the fingernail or toenail, which is why it doesn't complain too much about having rivets hammered in to hold steel shoes in place. But these are matters of detail, variations on a basic vertebrate plan that remains the same.

All vertebrates, however different superficially, including snakes, rabbits, lizards, birds and indeed humans, have a skeleton adapted from the same basic plan. A legless snake has some tiny pieces of bone recognisable as ancient vestiges of former limbs. The key point is that evolutionary adaptations of the vertebrate design, for instance in the construction of a bird's wings, have not involved the invention of completely new bones but instead have been adaptations of existing bones. The elephant's trunk (thought to be a snorkel) does not contain a skeleton and its tusks are merely incisor teeth.

Similarity and difference are merely points of view. At first glance the vertebrates display extremes of diversity. Differences in form, size and environmental preference need not be laboured as one thinks of hippopotamus and parrot, tortoise and giraffe, dinosaur and snake, fish and crocodile, whale and human. All are vertebrates. For decades the study of evolution has focused on difference, inevitably, since that is the basis of classification. But beauty is only skin-deep. Beneath the surface of difference is an astonishing uniformity to all of life, not just vertebrate life. Also beneath the surface is an almost unbelievable complexity within the structure of even the simplest cell, still more so in the intricacy and accuracy of the molecules that perform life's functions. It is at this level that the uniformity and beauty of all of life are so deeply impressive. As the study of evolution leads back towards the origin of life it becomes necessary to avoid the distractions of difference and to retell the story as one of similarity instead, a similarity across the dimensions of life and throughout its existence that leads to the conclusion that life has never changed much. Not on this Earth.

THE WORKING PARTS

The step from anatomy to physiology sees a quantum leap in hidden similarity. Take breathing. At first glance a fish seems markedly different from air-breathing animals because it lives under water. In fact both groups breathe air, or more precisely extract oxygen from it. Air is about 21 per cent oxygen and 79 per cent nitrogen, with traces of other gases including carbon dioxide, which is important to plants. The only real difference is that air-breathing higher animals have lungs that extract oxygen directly from the air, while true fish do not. Fish have gills that serve the same ultimate purpose but extract dissolved oxygen from water. Whales, porpoises, turtles, crocodiles and other species that in evolutionary terms have returned to the water after an air-breathing existence continue to rely on lungs, so can drown if they do not surface regularly. They have not re-evolved gills, which underlines a fundamental truism about evolution in this context: it does not go into reverse. An equivalent organ can evolve again, but not by reversal of a previous loss. Recent studies by Michael Whiting on stick insect evolution suggest that genes necessary for wing formation have on occasions been switched off and later turned on again, but that is not the same as a limb changing its structure and then changing back.

The similarity between lungs and gills is still greater considering that lungs are lined with a film of liquid. Both lungs and gills provide surfaces at which dissolved oxygen is extracted from water and finds its way into the bloodstream. By the reverse route, carbon dioxide passes out of the blood and into the environment through either the gill or the lung.

Almost every comparison one makes in biological science rekindles the debate over similarity or difference. Is a rabbit similar to a bird or completely different? Is a gill similar to a lung? Is a bat's wing similar to an insect's wing? The recurring answer is that it depends on your point of view! If similarity is stressed rather than difference, the inevitable conclusion is that life is extremely ancient.

ENERGY

From physiology to biochemistry is another quantum jump. What do living animals and plants actually do with the oxygen they consume? Here the similarities become even more striking. All air-breathing animals, indeed all oxygen-dependent forms of life, depend on one fundamental chemical reaction to obtain energy for survival. That reaction is simply the combining of hydrogen and oxygen to make water. A simpler reaction is hard to find, yet none is more crucial to the existence of life. (We also use a tiny amount of oxygen for making special molecules such as adrenaline, dopamine and skin pigment.)

In the good old days when school chemistry was real science, a favourite demonstration was the explosion of a mixture of hydrogen and oxygen in a special container by means of an electric spark. Two atoms of hydrogen combined with one atom of oxygen to form water, H_2O, in the form of hot steam, but the memorable part of the experiment was always the bang and sheet of blue flame, pressing home the message that a great deal of energy was released in the reaction. Every now and then a building or some unfortunate person's home is tragically blown to pieces after a leak of domestic gas, mixed with air, is ignited. Burning domestic gas, rich in hydrocarbons, generates water in the same reaction, which is why a cold kettle over a gas flame sweats with condensation as beads of water run down and hiss on the stove. The same dramatic release of energy when hydrogen combines with oxygen is what powers aerobic life, but nature's trick is to release the energy gently.

Living things need energy to stay alive and move about. Even when apparently resting, living organisms are using energy continuously. They need to pump the heart, move the intestines, expand the lungs, manufacture proteins and perform a host of biochemical processes. While asleep or relaxing, an average adult human's energy consumption is about 5 kilojoules per minute. Food labels commonly state the energy content, in either kilojoules or kilocalories (multiply kilocalories by 4.2 to get kilojoules). These numbers are of real interest to anyone wanting to control their weight (which is not an endorsement of fashion, but of health). A tub of yoghurt may be labelled 'energy 300 kilojoules per 100 millilitre serving'. That's just 60 minutes' energy supply for a person doing very little and using 5 kilojoules per minute. For a swimmer using, say, 30 kilojoules per minute, it would supply 10 minutes' worth of energy. The energy is released by the same familiar reaction of combining hydrogen with oxygen. The hydrogen is extracted from the food with the assistance of enzymes and the oxygen comes from the air, which is why breathing is necessary (anaerobic creatures also exist but their biochemistry is remarkably similar – see later).

The real trick of aerobic life is to combine the hydrogen and oxygen without the equivalent of a devastating gas explosion, otherwise one deep breath could be rather nasty. 'A flash of inspiration' would become standard, if colourful, obituary-speak and the surface of the Earth would be pock-marked with little craters where people had their bright ideas. 'Getting steamed up' would take on an entirely new meaning.

The combination of hydrogen with oxygen in living cells produces the same amount of water and releases the same amount of energy as if the mixture were burned under the kitchen kettle, but much of the energy is channelled into dozens of smaller reactions that handle it very briefly. Ultimately the energy is used to manufacture a chemical known as ATP (Adenosine TriPhosphate), which provides the immediate energy for muscular contraction and many other activities of life, such as growth. ATP is sometimes described as a high-energy compound or as having a high-energy bond, but that is inaccurate. It is the balance, or equilibrium, of the reaction that enables the transfer of a lot of energy. ATP is made as required rather than stored, but that's another story (Chapter 5).

Fascinating, but where are these observations leading? They highlight how seemingly diverse forms of life have an underlying unity because they are in fact related. Animals utterly different in appearance are very similar in physiology and practically identical in biochemistry. Familiar anatomical patterns such as the vertebrate skeleton lead to the belief that different species are related. Physiological similarities strengthen that belief. Biochemical similarities make it a certainty. Next we shall see how still more detailed examination of life's molecules, through their physics and the ways they are synthesised, allows relatedness to be measured precisely and expressed in numbers.

ANCESTRY

When people acknowledge that they are related, they know they had a common ancestor. Irrespective of how unrelated any two humans appear to be, they still had a common ancestor, who may have lived much more recently than they might at first guess. The number of ancestors every person unavoidably had doubles with each generation. Everyone had 2 parents, 4 grandparents, 8 great-grandparents, 16, 32, 64, 128, 256, 512, 1024... and so on, and the calculation is absolutely inescapable. The power of this doubling series is startling, amounting to over a million ancestors (1 048 576 to be precise; see Chapter 8) just 20 generations ago, say 400 or 500 years – and continuing to double every 20 years or so back.

This calculation is not to be confused with population growth. However gradually a population may have grown, or not grown or even shrunk, any individual unavoidably had 1 048 576 ancestors the twentieth generation back, or 2 097 150 ancestors if all the generations are summed. Some ancestors are likely to have put in more than one appearance in the family tree and if you trace back far enough, most of them would have done. Anyone who could trace all their ancestors back to the fifteenth century (which nobody can) would see that they are related to over a million people living at that time. Until the advent of mass international transport in recent decades, countries (in the loose geographical sense) had populations that increased gradually in numbers but generally did not move around much. If a person's country of ancestry had a population of about a million in the fifteenth century (and leaving aside the impact of migration) they are now related to just about everyone in the country! The same would apply to their partner. The pair would certainly have had a number of common ancestors, a term that, unless qualified, always implies the most recent. When this simple but inescapable arithmetic is explained, it often leads people to the sudden realisation that they are considerably more inbred than they thought, which goes a long way to explaining why ethnic groups may possess easily recognisable characteristics.

Sister and brother may have had a common parent a year ago or a decade ago. In the case of two seemingly 'unrelated' people originating in the same country, the common ancestor may have lived, say, 200 years ago. The common ancestor of two people of very different ethnic backgrounds could have lived much earlier, perhaps tens of thousands of years ago, but they had a common ancestor nevertheless.

BEYOND HUMANS

Ancestry is an easy concept to accept when confined to the human species or any single species in which interbreeding happens without a problem. The next step in the argument is to consider the common ancestry of different species that cannot interbreed. Incidentally, the meaning of the word species can be a slightly moving target, but the key part of the definition for this discussion is that different species do not naturally interbreed *and produce fertile offspring* (the last part has to be added otherwise mules and hinnies mess up the definition).

Consider the ancestry of humans and apes. Both are mammalian vertebrates with so many similarities that they are grouped together with lemurs and monkeys as 'primates'. Humans and apes are not the same species but are evidently related through a common ancestor, perhaps not very different from a modern ape, who lived a few million years ago. A new problem starts to emerge as one jumps from comparing different humans to comparing different species: trying to deduce exactly what the ancestor looked like becomes increasingly difficult.

It's easy to say that all humans are related and fairly easy to say that humans and apes are related, on the basis of what they have in common. The same sort of logic leads to the observation that sheep and goats have a lot in common. Presumably they are descended from a relatively recent ancestor with horns, a woolly coat and a taste for grass. Tigers and leopards are related to each other in having well-known cat-like features and an ability to chase somewhat faster-moving prey than grass. Within each group the similarities are obvious, but when the sheep group and the cat group are compared it's the differences that stand out. Nevertheless, numerous similarities remain such as the vertebrate skeleton, blood and circulation, mammalian reproduction, eye structure and so on. The really difficult problem is to visualise exactly what the single common ancestor of the sheep and cat families looked like, beyond the fairly obvious expectation that it was a four-legged vertebrate mammal. Since these two rather different families are descended from it, the common ancestor must have lived earlier than both the ancestral cat and the ancestral sheep–goat.

The so-called evolutionary tree (see Figure 2.1) has been constructed by this sort of reasoning. Perhaps it would be better to say that the tree *is being* constructed as new findings continue to improve our understanding of relationships. Earlier versions of the tree can now be subtly out of fashion. The principle of the evolutionary tree is that any two species (to be precise, two individuals of different species) must have had a common ancestor. The position of that common ancestor is a branch point on the tree, but that doesn't mean that anyone can properly picture the ancestor. The more two species differ, the nearer to the root of the tree (the more generations ago) will be the point where they branched from their common ancestor.

An elementary but common mistake is to imagine that a species living today is the ancestor of another. Humans are not descended from today's apes, although the two had a common ancestor. When people talk about humans being descended from apes, or of apes being our closest relatives, they mean that the species with which we shared an ancestor most recently, as far as we know, was ape-like. This is an important distinction, because just as one cannot be certain what the cat–sheep ancestor was like, the exact nature of the ape–human ancestor is also debatable, even if current evidence suggests that it was more ape-like than human.

It's very easy to get things wrong. Last century when the evidence was scarce, a great many people got into an awful lot of trouble

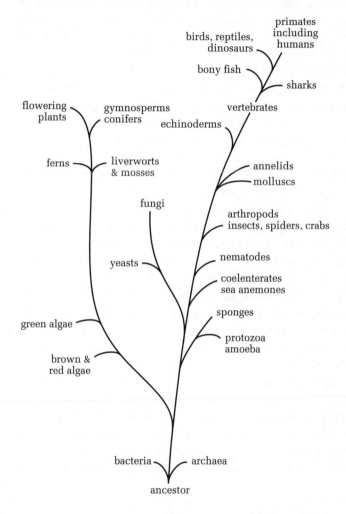

Figure 2.1   A typical representation of the evolutionary tree

following a fraudulent trail towards a preconception of a human brain on an ape's shoulders. Scientists are not supposed to yield unto temptation, but so intense was the pressure to find an imaginary half-ape, half-human ancestor that eventually some obliging individual manufactured one. The Piltdown Man forgery of 1908–1912 remains one of the great scientific whodunnits and has inspired countless theories. The fake discovery was set up in a gravel pit near Piltdown in southern England. Planted fragments of cranium were genuine centuries-old human and easy to extrapolate, correctly, to a large brainbox. Placed with them was a similarly old jaw fragment from an orang-utan, cunningly broken to remove tell-tale signs of its true provenance and the fact that it would not have articulated properly against the upper skull. The teeth were reshaped by filing and the whole invention was darkened with chemical stains to look the right sort of age. To help the illusion, the discovery site was seeded with a few stone and bone tools and a sprinkling of genuine fossil animal teeth. The finder of this little surprise was expected to exclaim, 'Aha! Ape jaw, human brainbox. The missing link between ape and human!' Which they did. How wonderfully proper it was for the world's first gentleman to have lived near London, a bus ride from Darwin's back garden.

Unfortunately for the forger, ideas about the origin of humans changed and the skull became scientifically ridiculous. The openness of science ensures that dubious results will collapse sooner or later. Concerted analysis between 1949 and 1953 with new techniques that the forger had not foreseen, principally the measurement of fluorine absorbed from groundwater, finally proved that the cranium and jaw never belonged to each other.

EXTINCT LIFE

Looking much further back and trying to imagine the shape and appearance of the first mammal or the first vertebrate lands us truly in the realms of conjecture. The temptation when trying to envisage an unknown ancestor is to put it together from the features of known species and known fossils. Likewise, when an evolutionary home has to be found for an extinct fossil, the urge is to place it at a key branch point in the evolutionary tree, as the common ancestor of two later families. Unfortunately, the probability of finding a true ancestral fossil is so small that a fossil is much more likely to represent a small twig of the tree. Both the fossil and the modern species apparently related to it are derived from a still earlier ancestor.

The relationship of reptiles, birds and dinosaurs is worth looking into as a prime example of the difficulties involved in arranging the evolutionary tree. Many characteristics of birds and reptiles, including internal anatomy such as the way the blood vessels are connected to the heart, support the idea that both families are closely related (the word 'family' is used loosely in this book). This idea has been around at least since the 1859 discovery of the 150-million-year-old fossil *Archaeopteryx lithographica* ('ancient wing imprinted in stone') in a quarry at Solnhofen, Bavaria. Other specimens have been found since. *Archaeopteryx* had feathers and other bird-like features in addition to reptilian characteristics and teeth. That doesn't make it the 'missing link' between birds and reptiles, although for a long time it was popularly regarded as just that. It almost certainly wasn't.

Archaeopteryx was a very early bird, to be sure, but there is no proof that it was the one and only ancestral bird, the very first reptile to sprout feathers and take to the air. For a while it was the one and only candidate, which is not the same thing. Rival candidates have been unearthed in recent years, such as *Confuciusornis sanctus* (found in China, need one add) which has a beak closer to those of modern birds, tending to sideline *Archaeopteryx*; and *Protarchaeopteryx robusta*, also found in China, which is in the early days of analysis but has some features a little more primitive than *Archaeopteryx*. The pattern of this kind of research is that ever better contenders continually come to light, straightening the evolutionary path and displacing previous candidates into little side branches.

A popular theory is that modern birds are descended from small dinosaurs, or ancestors of dinosaurs, that adopted the habit of walking on their hindlegs, evolved feathers (perhaps not in that order) and developed the ability to fly. Indeed, many people say that birds are simply flying dinosaurs. More recently, however, the reverse has been proposed – that at least some dinosaurs (in the Maniraptora group, having 'hands') might have been derived from birds that lost the power of flight. Either way, did birds develop flight by taking off from the ground and landing in the trees, or by launching from the trees and landing on the ground, perhaps via other trees?

There are a number of separate issues here, including bipedalism and flight.

Let's look first at bipedalism. Envisage a species adopting a tendency to use its forelimbs a little differently from its hindlegs, as a cat pounces mainly with the power of its hindlegs but grips prey with its front paws. Such a tendency grew from small beginnings. The advantage to the dinosaur of its first tentative bipedalism may have been additional height to reach tall plants, freedom to use its forelegs for food handling or its head remaining above the swamp, but whatever it was, this leading evolutionary change was selected for.

It's extremely important not to fall into the trap of thinking that some mysterious desire for bipedalism was acquired first and imprinted into the genes for subsequent generations. Evolution doesn't work that way. Mutations (changes) happen first and many do not matter at all; they are called neutral. Occasionally a mutation conveys some small advantage and suits a slight change of structure or behaviour that in turn may suit a slightly changed environment. A major shift of behaviour and structure leading to bipedalism would represent a vast accumulation of mutations over many millions of years. The more beneficial a collection of mutations, the more rapidly they would spread through the population and become 'fixed'.

Evidently a line of dinosaurs accumulated mutations favouring a bipedal habit that conveyed some advantage, but this is not a trivial process. Consider the global reorganisation of the animal necessary for it to happen. Shrinking the front legs of a quadruped does not make it into a biped, it would fall on its chin. A nose-to-tail remodelling of the animal is required. The centre of gravity needs to shift hindwards to restore balance, otherwise standing on the hindlegs is a great effort and wastes energy. A heavy tail is desirable to aid this shift of balance, tilting the stern down and the front up. This imposes extra loads on the hindlegs and hip joints, which need to be re-engineered to bear twice the weight. The brain and nervous system have to be rewired to control two-legged walking, which is not simply half of four-legged walking but demands a completely different form of balance. (Four-legged walking and running are surprisingly complex and were misunderstood before the inspired studies of the artist George Stubbs in the eighteenth century and the photographic studies by Eadweard Muybridge in the nineteenth century.) The newly liberated forelimbs require an independent system of control for their non-walking activities. The head needs to be tilted down at a lower angle. A longer neck can be supported more easily.

These are merely a few of the glaringly obvious areas of reconstruction needed before the old quadruped dinosaur can be relaunched as the new biped model. The positive benefits of a reduction in forelimb size now become self-evident. A glance at the *Tyrannosaurus rex* skeleton shows that its weight was logically distributed for a bipedal habit thanks to a massive tail, heavy hindlegs and a tapered upper body with tiny lightweight forelimbs. It was not front heavy and kept its chin off the ground.

Bipedalism is an evolutionary development in its own right, having positive advantages irrelevant to flight. Evolutionary history has repeated itself in the modern kangaroo: heavy tail, massive hindlegs and hips, small upper body, lightweight forelimbs and a completely reprogrammed gait, with the tail recruited as a walking accessory and assisting support like a tripod.

Now let's look at the evolution of flight. A completely different pathway can be envisaged in which bipedalism was secondary, not primary. The two main theories of bird flight are that it originated either from gliding down out of tree tops or through launching from the ground into the air. Consider first the value of gliding to animals that survive by catching prey and avoiding becoming prey. The ability to climb trees or cliffs would have been beneficial either way, as would the ability to take the quickest route down by jumping. The laws of motion dictate that hitting the ground from twice the height dissipates twice the kinetic energy, or loosely one hits twice as hard, but the prospect of injury increases much more steeply. What's needed is a way of slowing descent. Better still would be some form of steering to target fleeing prey or another tree that might be a better escape route than the ground straight down. When an animal leaps off a tree branch it has forward motion and some degree of aim, but the front end starts to drop before the back end lets go, so the body tumbles steadily and would tend to hit the ground head first (it would continue to rotate through head up, but somebody's law dictates crashing at the worst moment). The same happens when a swimmer dives off a high board or a car is driven off the end of a jetty (those beloved film stunts where a car jumps the gap in a bridge are carefully engineered so that the front wheels are flicked upwards by a ramp, which instantly collapses before the back wheels pass over). The anatomy of front legs, collar-bone, neck and head is nowhere near as well suited to hitting the ground from a great height as are the back legs.

Not forgetting that evolution is blind to need, it never passes up a good offer either. Feathers were an offer it could not possibly refuse, solving in one go all the problems of the previous paragraph. Feathers are mainly the protein keratin. The gap in the fossil record as to how they originated and evolved is rather large, as opponents of evolution are ever pleased to point out. This doesn't mean that feathers did not evolve, since obviously they did. The fossil gap is not surprising because the advantage of feathers would have been so immense and so rapidly adopted that fossils of transitional species, with feathers only partly evolved, should be exceedingly scarce.

The fossil evidence includes a recently discovered Chinese specimen considered to be a dinosaur with feathers on its body and on all four legs. Significantly, this dinosaur is bird sized, its feathers are not very different from modern flight feathers except for being symmetrical, and it has feathers on its tail. *Archaeopteryx* itself had signs of possible hindlimb feathers according to Nick Longrich, as did other fossil birds of the period. By increasing drag, feathers slow descent. As air brakes on the front limbs they can be used to control pitch in the glide, counteracting a head-down rotation and so helping to keep the landing target in view and the hindlegs lowered for touchdown. By manipulating its front limbs differentially, the gliding animal can bank and turn. By tilting both wings up and optionally deploying feathered hindlimbs and the tail as extra drag, the animal can stall (discard lift) as it touches down.

Once the evolutionary switch down this pathway was thrown, a number of developments would inevitably have been selected. A small step from controlled gliding by movement of the wings is a little more movement, adding propulsion. Specialisation of the hindlegs for impact, and for seizing prey, would have seen strength selected for and feathers unnecessary there, while the wings developed in the direction of lightness, fine control, more efficient feather technology and powerful musculature. Bipedalism is inevitable by this route, but through a different set of evolutionary pressures from those leading directly to bipedal dinosaurs.

The opposite idea is that primitive birds took off from the ground, possibly after a phase of feather-boosted running in pursuit of prey. Birds of this model might also have glided initially, just above the ground where a special air-cushion effect within half a wingspan of the ground (called ground effect) makes it a little easier. But there are also problems, including a greater learning curve (in reality, hard-wiring of the nervous system, see Chapter 10), a higher power requirement for climbing out, a faster wing beat, and other obvious disadvantages compared with a controlled plummet from modest altitude. Considerable debate has gone into whether *Archaeopteryx* and its cousins could have taken off from the ground or must have launched themselves from treetops. Calculations suggest that *Archaeopteryx* could not have reached take-off velocity solely by running, although wing action might have helped it up to speed. Evidence from claw shape and estimates of the quantity of flight muscle point more to a habit of climbing trees and gliding back to the ground.

The bipedalism of dinosaurs is not that of birds, for which it was not a matter of standing taller or catching prey with the forelegs, but of moving faster – and incidentally in birds has led to the opposite behaviour of carrying cargo with the hindlimbs. The principle of flight, powered or unpowered, is that the body is suspended below the wings, critically balanced between fore and aft. Light aircraft have fuel tanks positioned in the wings partly so that the centre of gravity stays in the same place when fuel is consumed. The very last thing a bird wants is a heavy dinosaur tail, which would tip it nose up and cause it to stall. Actually it wouldn't get the chance to stall, it would never get off the ground. The bird has a far better use for the tail as a rudder and aerodynamic control surface for lateral stability, for which it needs to be least of all like a solid dinosaur tail but instead as light and expandable as possible – feathers. Of course, many more adaptations than the development of feathers and a shift in the centre of gravity were necessary to equip vertebrate life for flight. Many of these features have been reviewed in detail by Pat Shipman (see Appendix).

The evolutionary path to two-legged dinosaurs could not have led on to birds without reversals on a scale that do not happen. Evolutionary precedent does not support the idea of shrunken, remodelled limbs being restored to their original structure. Even less is there precedent for a regressed limb re-evolving into something more complex than it was in the first place. On the other hand, there is good evidence that the bird wing is an adaptation of the full-sized and fully functional vertebrate forelimb. With or without feathers, the dinosaur's degenerate forelimb could not have helped it into the air. The tag 'degenerate' is in any case unjustified, because those dinosaurs succeeded in their niche and small forelimbs were clearly beneficial. Evolutionary degeneration is not the trivial loss or diminution of a limb. The shift to bipedalism is a global reprogramming of the genetic expression of the entire organism, with all the positive changes accruing, the re-engineering of the forelimb being a small part of the overall process.

The converse idea of a branch of the dinosaurs developing from flightless birds has the serious flaw that the bird would need to redevelop a massive tail after having replaced it with a small one with feathers. Conversion of the bird's clawed wing back into a clawed arm is also somewhat contrived. While it is true that lineages often tend to evolve towards larger forms rather than smaller, the evolutionary fate of flightless birds is well demonstrated by species walking about today or only recently extinct, including the emu, ostrich, dodo and the enormous moa. Their wings are either miniscule vestiges or not even that, and have certainly never redeveloped.

The debate about the origin of birds has been mentioned to make deeper points. One is that at the dawn of the twenty-first century,

150 years after Darwin, the debate is as hot as ever over the true relationships within one of the great divisions of life: birds, reptiles and dinosaurs. Another (to be discussed in later chapters) is that if it is so difficult to piece together evolution retrospectively, it is impossible to project it forward. Despite the evidence from fossils, anatomical studies and modern molecular techniques, it remains uncertain how these families were originally related. Yet all are vertebrates living within the past 300 million years, a fraction of the 4000 million years or so of Earth's existence. If it's that difficult to be certain of relationships in the last 7 per cent of the timeframe available for life on Earth, where the richest collection of fossil evidence is to be found and much of the evidence is still walking and flying around, then we are in for a long haul to see back the other 93 per cent.

So far only vertebrates have been mentioned and it isn't difficult to accept that all vertebrates are related by an underlying unity of structure and function. They are descended from a hypothetical first vertebrate, the shape and appearance of which cannot even be guessed. But how far can one push the idea of the relatedness of life? Are vertebrates in turn related to the invertebrates, such as insects, sea urchins, worms? Could there even be a direct connection between the two great divisions of life, animals and plants?

INVERTEBRATES

Most biologists agree that all forms of life are related by common, if extremely remote, ancestry. Sometimes the relationship is considerably more obscure than it appears to the non-biologist, but the relationship is present nevertheless. Compare vertebrates and insects. Vertebrates walk around on four or two legs usually; they may have wings; they generally have a pair of eyes. Insects are much the same, are they not? Does it matter that adult insects walk on six legs rather than four? And don't many of them also have wings and a pair of eyes? Just like vertebrates?

True. But these similarities are extremely superficial and conceal great differences between vertebrates and insects. The legs do the same job of moving the animal, but are built completely differently. Vertebrates have an internal skeleton made of calcium-rich bone surrounded by muscles, blood vessels and a skin wrapped around the outside. Insects have an external, not internal, skeleton constructed of carbohydrate (chains of sugar molecules), protein and other components with muscles attached to the inside, but no bones. Each of these two alternative body plans has advantages and disadvantages. The external skeleton of the insect, which is known as arthropod ('jointed foot'), is in many ways more protective but introduces the problem of its having to be discarded by the growing animal from time to time through moulting, making the animal vulnerable to predation while this is happening.

The same extent of difference is seen in the wings, which are so utterly different in concept and construction between birds and insects that there is nothing to compare, except that they support flight. Birds and bats are both vertebrates, yet the difference between their wing structures is again so fundamental that they can scarcely be compared, the bird relying on feathers and the bat, a mammal, relying on a webbing of skin and tissue.

The eyes are also completely different, with a single lens projecting a single image onto the retina of the vertebrate, but many light-sensitive detectors making up the compound eye of the insect.

Vertebrates can be quite diverse but invertebrates are far more so, with numerous, seemingly different body plans. Worms are not merely worms. Nematode worms are anatomically quite different from earthworms, which are different from parasitic tapeworms and again from flatworms. The variety of life forms seen among invertebrates is astronomical. Some of the fattest reference books are attempts to catalogue invertebrates, notably the many thick volumes by the legendary Libbie Henrietta Hyman.

On the other hand, invertebrates have far more in common with each other than is superficially obvious. Butterflies are remarkably similar to crabs, although crabs are not insects; but both are arthropods with a common ancestor. Jellyfish have much in common with sea anemones. Despite their spectacular differences, all invertebrates are closely related.

ALL LIFE IS RELATED

The opening question was whether life on Earth is traceable to a single beginning, because if so, only a single date needs to be assigned to that beginning. The evidence is plain that vertebrates, whether of the water, land or air, are a single family despite the numerous guises in which they appear. Similarly, the still more diverse, often bizarre invertebrates are related to each other. Superficially there is nothing in common between vertebrates and insects. Yet the truth is that they have enough in common to be unquestionably related. Both are constructed of cells having similar internal components including a nucleus, mitochondria and ribosomes, both have considerable biochemistry in common, use the same pathways of metabolism, have the same kind of genes made of DNA (deoxyribonucleic acid) and use the same machinery to translate genetic information into proteins. As the enquiry is pursued into single-celled animals, algae, fungi, bacteria and plants, the conclusion continues to strengthen: all are related and unified by a common style of biochemistry. They have a common source.

Until relatively recent times the source of life was beyond hope of comprehension. The creation of life was literally supernatural and outside the realms of science. The application of scientific enquiry, however, reduces the problem to two very simple questions. First, did modern complex life evolve from a vastly less complex single-celled ancestor through a process of change? If the answer to the first question is 'yes', then all manner of previously intractable questions about how creatures as complex and special as humans could have come into existence are no longer questions, they are answers, observations of historical fact.

The second question fits chronologically before the first: how did life for the first time reach the stage of complexity of the simplest plausible single-celled organism, simpler still than the sophisticated bacteria that now account for most of the living organisms on Earth? The answer to this question, stripped of all but scientific logic, has to be rather mundane. Since life is chemistry and nothing but chemistry, the chemistry happened. Chemistry is solely the consequence of atoms and molecules coming into collision. They reacted and combined in ways that might seem absurdly unlikely and may have taken a very long time by human standards, but there is no future in arguing that something cannot happen after it already has. No branch of science can suggest any alternative. The main difficulty with this view is that the achievement of the first single-celled stage of life seems inherently less likely and more time-consuming than life's later divergence into multifarious species.

There is no evidence to refute the theory that all life on Earth is related. If that is so, then all present life can trace its ancestry back to the same primordial individual, presumably a single-celled organism, the first truly successful living thing. Earlier false starts to life probably happened and fizzled out, so it is not accurate to say that life can be traced back to the first cell, but rather to a cell whose descendants have replicated successfully from that day to this. Hypothetically the theory could be disproved or at least seriously weakened by finding a form of life that has nothing in common with others, such as a completely different way of obtaining energy, a different cellular structure, a machine different from the ribosome for assembling proteins, a genetic information store different from DNA, or a completely different genetic code. Variations do abound in all of these features, but nothing sufficient to overturn the theory of a common origin. As with any good theory, it constantly invites disproof but stands until that happens.

Life at the level of cells and their internal components is so overwhelmingly uniform that it would be very difficult to argue against a single common origin, but the next unanswered question is, when did this origin happen?

Dating the Ancestors

*When a timescale is added to relatedness,
life is found to be very old*

The idea of different species being related to a common ancestor is so easily accepted nowadays that it shakes one to realise that the idea has been popular since no more than a century and a half ago. That could have been in the lifetime of our great-grandparents. When Charles Darwin published *On the Origin of Species by Means of Natural Selection or the Preservation of Favoured Races in the Struggle for Life* in 1859, to the anticipated hostile reception, he had done his homework. He made his observations over at least two decades, founded a solid theory on them and published it. The onus was on the disbelievers to disprove the theory – that is the way of science. People undoubtedly had similar notions long before Darwin and some voiced them, but to do so was to risk being branded a social outcast, a lunatic, a heretic, an agent of the devil.

Alfred Russel Wallace had more than notions at about the same time, he had very much the right ideas and used to write to Darwin about them. Wallace published a learned paper on his ideas a year before Darwin's great work appeared, but Darwin receives most of the credit because his work was more complete, he had more evidence from his world travels and his ideas were of longer standing. 'In science,' said Sir William Osler (1838–1923), 'credit goes to the man [sic] who convinces the world, not to the man to whom the idea first occurs,' as we have seen before.

Another interesting early contributor to the debate, Robert Chambers, was a man of humble background. He rose from poverty as the son of a bankrupted weaver to become a popular author on a variety of topics, such as the book *Traditions of Edinburgh*, the city in which he and his brother William built a successful publishing business. Robert would probably have become Lord Provost of Edinburgh if he had not written *Vestiges of the Natural History of Creation* in 1844, which predictably did little for Creation. Even the staid *Scientific American* magazine took a swipe at the 'nonsense' of *Vestiges* in 1852. The book was a runaway success nevertheless, selling in eleven editions in three languages in Chambers' lifetime and other editions since. To say that it was authored anonymously would understate the highly elaborate precautions for concealing Chambers' name. In his own words, 'the author's name is retained in its original obscurity, and, in all probability, will never be generally known'. But someone soon found out and for his heresy Chambers was pressured into withdrawing his candidature for high office.

Johann Wolfgang von Goethe (1749–1832), German poet and amateur scientist, proposed a theory of evolution 80 years earlier than Darwin, in between learning several languages, writing *Faust*, studying art, chemistry, theology, anatomy, architecture and topping it off with a law degree.

Earlier still, naturalist Georges Louis Leclerc, Comte de Buffon (1707–1788), clearly disbelieved the ecclesiastical line about both the age – or youth – of the Earth (about 6000 years according to archbishop Ussher) and the creation of life. Buffon's contribution to the debate is mentioned in Chapter 1. He published prolifically, including his life's work the *Histoire Naturelle, Générale et Particulière* in 44 volumes, but in eighteenth-century France he could not write anything that displeased the censors. As a result, it is often said that he never completely embraced the theory of evolution, but a paragraph slipped into his *Theory of the Earth* eliminates all doubt: 'It can be assumed that all animals descend from a single living being which in the course of time, by perfection and degeneration, produced all other animal forms.' What he meant by degeneration is open to debate, but 'all animals descend from a single living being' clearly refers to the idea of common ancestry.

Those are just a few of the perceptive early contributors to the ideas of relationship and common ancestry, on which they were broadly correct. What they lacked was a realistic absolute dating system.

DATING ANCESTRY

For centuries before Darwin, people had been discovering fossils that seemed to be non-existent animals and plants. Every possible fanciful notion as to the meaning of fossils was dreamed up (such as they were planted to confuse us) except the blindingly obvious – fossils are dead plants or animals (or imprints of them) that have been preserved in the earth since they died. Since their kind often no longer existed, life had since changed. Once this was finally realised, it became possible both to chart the changes in life during more recent evolutionary history and to apply a dating scale to those changes, if a dating device could be found.

An indirect approach to dating is to compare modern living species and work out how long ago they had a common ancestor. This method means theorising what the common ancestor was like and defining the time when such an ancestor may have existed, and therefore depends for calibration on the direct method of dating the fossils. The crux of the problem is the need to depend on comparisons between species alive today while trying to deduce something about a hypothetical ancestor that no longer exists. How the ancestor looked is at the very least uncertain and if it dates back a long way, say to the origin of vertebrates, we have only the sketchiest notion.

The information obtainable from fossils is limited because normally only the hard parts are preserved. If the fossil is of a bony skeleton, it has to be fleshed out according to what people think the outer body probably looked like. Some highly authoritative photofits of Piltdown Man (Chapter 2) were cobbled together – considering that he never existed – complete with finger and toe details, based on small fragments of skull. Conversely, animals with a hard outer structure such as sea urchins may have left only an impression of the outer surface and so the nature of the soft parts inside is a matter of interpretation. From small clues such as the shape and wear of fossil teeth, scientists have to reconstruct what an animal ate, what it looked like and how it lived. Some fossils are more complete than others, with maybe an outline of a fish as well as its skeleton, or an impression of feathers with the skeleton of a bird. Trees are found petrified throughout the thickness of the trunk with the internal structure discernible. Very occasionally the internal organs of fossils have been preserved when fossilisation happened rapidly, as in a renowned Brazilian deposit. A technique devised recently has revealed impressions of soft tissue in fossils preserved in volcanic ash, including molluscs and crustaceans, by imaging a succession of polished sections.

Provided that it is treated with caution, the fossil record provides a broad overview of the increasing complexity and development of life over hundreds of millions of years. But still the crucial problem in interpreting the fossil record is a reliable timescale.

DATES FROM ROCKS

Fossils are dated by asking a geologist the age of the rocks in which they are found (and the geologist, it is blasphemously said, dates the rock by asking a palaeontologist the age of the fossils!). Fossils are formed usually as a result of animals or plants becoming buried in underwater sediments. The sediment builds up over a period of time that can be very long in a deep ocean. The most recent layers containing the youngest fossils are nearer the top, or what was originally the top if the strata have tilted through geological upheaval. Usually the sedimentary rock is dated slightly indirectly by dating the igneous (solidified volcanic) rock adjacent to it. This is done from radio-isotope measurements as mentioned in Chapter 1, but it is not easy, either to do or to interpret. The principle is that when a rock is first formed, traces of radioactive isotopes are present in certain ratios. Different isotopes have different but precisely known half-lives and they decay naturally over time. Today's ratio of the amount of an isotope compared with its decay products indicates how long this has been happening. Solidified volcanic rocks do not themselves contain fossils, but help to date fossil-bearing strata lying above or below.

Uranium 238 has a half-life of 4.47 billion years or about the age of the Earth and is therefore capable of giving dates at the longer end of the geological timescale, from about 50 million years ago back to the year dot. The choice of isotopes for measurement depends on what isotopes are present in the rock. If the uranium/ lead method is unsuitable, alternatives include potassium 40, which decays partly to argon 40 (half-life 1.25 billion years), or rubidium 87, which decays to strontium 87 (half-life 48.8 billion years).

Fine detail can be seen inside sedimentary rocks. In some it is possible to see layers, similar to tree rings, that correspond to seasonal events such as winter floods, and so to calibrate the annual rate of rock growth. Maximum and minimum limits can be set for sedimentation rates to form a particular type of rock. It is possible to put at least relative dates on the different layers even if the absolute age is more difficult to calculate. Fossils found in different layers can be shown to have lived millions of years apart.

DATES FROM PROTEINS AND DNA

With modern technology the inspired guesswork of relating species by appearance and physiology can now be augmented by accurate measurements of protein and DNA differences. Previously, fish and whales had nebulous similarities such as a vertebrate skeleton and a swimming habit, together with crucial differences such as one being an egg-layer and the other a mammal. Protein and DNA sequences make it possible to put numbers into the relationships and to say that two species are perhaps 30 per cent or 90 per cent different.

Either protein or DNA can be used because in a sense they carry the same information. A protein is a sequence of different units (amino acids) forming a long chain. DNA is the sequence of information for making that protein and is also a long chain, but made of different chemical units (nucleotides, containing bases). The more closely two species are related, the more similar are their protein sequences or their DNA sequences.

Proteins are the most important constituents of the cell, therefore of the body, and there are thousands of different proteins. Virtually every part of the cell and every part of the body is either made of protein or, if not, is made by the action of enzymes, which are catalytic proteins. Every feature of life is either constructed of protein, constructed by proteins acting as enzymes, or controlled by proteins such as protein hormones. The instructions (genes) for making those proteins are the instructions for duplicating the entire living organism.

The components of protein chains are known as amino acids because they have an amino group (H_2N-) at one end and an acid group ($-COOH$) at the other end. The usual kitset of amino acids for making proteins contains only 20 different kinds, with names such as valine, histidine, leucine. What makes one protein different from another is the total number of amino acids and the exact order, or sequence, in which they are joined together.

Obviously the manufacture of a large and complex molecule such as a protein, with maybe 40 or 200 amino acids joined together in exactly the correct sequence, is a complicated operation. Yet the cell manages it. The instructions are held in genes.

Genes consist of DNA. Genes have two jobs to do, one to provide the information for assembling proteins, the other to pass on this information from generation to generation when living cells reproduce. An important point to bear in mind is that evolution actually happens in DNA when bases become changed (mutated), but the effect of the change is felt in the protein if the mutation causes a different amino acid to be incorporated.

The genome of an organism – its complete set of genes – contains the complete know-how for assembling the organism. This has to be so, because in the process of reproduction the DNA in (mainly) the nucleus of a single fertilised egg cell is all that is required to control growth from that one cell into a complete living organism containing, in a human, about 1000 billion cells. The amount or length of DNA a species has, copied into its every cell, is not a good guide to the number of genes present, as much of the DNA is not used. Sequencing the human genome has shown it to contain fewer than 30 000 genes (as few as 24 500 according to some reports), a surprisingly small number when the microscopic worm *Caenorhabditis elegans*, a millimetre long with fewer than 1000 cells, has about 19 000 genes. This underlines how the complexity of life lies mainly inside the cell, whereas species differences reflect how different body plans are organised.

Although the genes contain all the information to replicate existing life, they are insufficient to create it on their own. A set of genes given all the small chemicals found in life could not generate a living cell, because the interpretation of the genetic instructions requires an already living cell. Genes can instruct the synthesis of proteins, but only if the amino acids and the cell's constructional machinery, in the form of ribosomes and numerous enzymes, already exist. Genes control the growth of cell membranes and the division of the cell into two, provided that the cell already exists to divide. Genes control the production of proteins needed by cell components such as mitochondria and lysosomes, which must already exist to receive and incorporate them. No amount of gene activity will cause a random mixture of proteins, membranes and biochemicals to fall together as a cell.

Just as a protein is a continuous sequence of amino acids, a gene is a continuous sequence of DNA. The two correspond such that the sequence of units in the DNA specifies the sequence of amino acids in the final protein. The units in DNA ought to be called nucleotides, which differ according the chemical base present (for the record: cytosine, adenine, thymine or guanine). For ease people often refer to the sequence of bases, and save ink by abbreviating them to C, A, T and G. DNA is constructed of not one sequence strand but two, twined around each other to form the well-known double helix. The two strands are opposites in the sense that the base A in one strand always fits or 'pairs' with the base T in the other strand, and in a similar way the base C always pairs with G.

For the time being it can be said that one of the strands of the double helix contains the genes, the other does not (but in places they swap). It needs a continuous sequence of three bases (called one codon) to carry enough information to instruct the cell to incorporate one particular amino acid from the choice of 20 into a protein. For example, the codon TTT means incorporate phenylalanine, GTG means incorporate valine (the complete translation table is provided in the appendix). A part of a gene that reads

TTT GTG AAC CAA ... ...

would be translated ultimately into a piece of protein with the amino acid sequence

phenylalanine – valine – asparagine – glutamine...

But suppose an individual was born with a mutation in the gene such that the first TTT had been changed to TTA. The protein would be built with a leucine in place of the usual phenylalanine (see the table), which is the sort of substitution that a protein might accept because these two amino acids have a degree of similarity, so the mutation might well be survivable. In this way individuals become different and species accumulate mutations and progressively change. The degrees of difference, the numbers of mutations and subtleties about their character, all add up to a measurement of relatedness. While the principle is easy, the interpretation is full of difficulties. In principle, the longer the time since two protein (or DNA) sequences had a common ancestor, the more differences between them.

NECESSITY FOR CHANGE

One of the paradoxes of life is that the duplication of DNA in preparation for cell division has to be exquisitely accurate, otherwise the process would fail, yet if DNA were perfectly conserved and replicated, change and evolution would be impossible. The human genome contains around 3 200 000 000 bases in sequence in each of the two strands of DNA (3.2 billion base pairs). Rather surprisingly, only around 3 per cent of this is conventional gene sequence, amounting to fewer than 30 000 genes. The remaining 97 per cent is far from written off yet as 'junk DNA', as evidence is coming to light that some of it may code for small but vital sequences of RNA (ribonucleic acid). Nevertheless, this vast amount of information has to be copied at each cell division. A single error in the DNA duplication process, or pre-existing damage passed on, could be so disruptive that the daughter cell is unable to survive. The cell dies, and with it the mutation. If that cell is a key part of the body structure, the organism is deficient and likely to die, and if the cell is in the reproductive germ line, inheritance of a line of ancestors comes to an end. On the other hand, if DNA never changed, life would never have progressed – in fact it probably would never have formed.

The equilibrium life has adapted to – unintentionally of course since this is pure chemistry – is that the error rate in the storage and duplication of DNA is incredibly low but not so perfect as to prevent change. Errors are so exceedingly rare that a population adopts a sustainable and non-lethal mutation perhaps every few million years in any one gene. If mutations happened much more frequently, life would be unstable and sexual reproduction (which requires the chromosomes of both parents to be pretty much alike) would be difficult. If mutations happened much less frequently, life would have been deprived of the opportunity to evolve and develop at the rate it has.

The process of error, mutation and evolution is subtly driven by the second law of thermodynamics, which states that entropy, or disorder, continues to increase in any system (Chapter 6). The randomness of collisions between molecules means that every now and then, the wrong base will be incorporated into DNA, equivalent to an increase in disorder. The likelihood of error is small for chemical reasons and the cell has cunning enzymes that minimise mistakes during replication and can even repair them afterwards, but a few do slip through. The second law of thermodynamics is often embraced simplistically to claim that because life is highly ordered, it could never have formed spontaneously; evolution could only lead to increased disorder and degeneration. The origin and evolution of life are thermodynamically impossible, it is said, therefore life must have been created supernaturally. On the contrary, by accumulating genetic changes life does not defy but obeys the laws of thermodynamics, which thereby work entirely in favour of evolution. Otherwise, humans would not be here today.

To turn these important principles into a dating device, a protein is chosen that is well understood and occurs commonly in different forms of life; many fit this description, conveniently, owing to the underlying uniformity of life. Small samples of the chosen protein are extracted from the two species to be compared and put into a piece of expensive technology that reads their amino acid sequences. When the two sequences are printed side by side, it is easy to count the number of differences.

At least, that's how it was done until a few years ago and sometimes still is. However, it just happens to be much easier and faster nowadays to sequence DNA rather than protein, and of course the DNA sequence can then be translated into the protein sequence, as in the example shown previously. The difficulty with the DNA method is finding the right piece of DNA to sequence, as much of it, including stretches of the DNA inside genes, is not actually translated into protein. Of the 3 200 000 000 base pairs in the human genome, the length coding for one small protein might be a mere 300 or so, or 0.00001 per cent. Nevertheless, there are methods that can be used.

THE TIME DIMENSION

If relatedness can be measured accurately by comparing protein sequences (or the same information obtained from DNA), all that's needed is to calibrate the number of differences versus time, and a new science of sequence dating is born. In future, simply count the sequence differences between two species, look at the calibration graph and read off how long ago they had a common ancestor. Right?

Wrong! It's a nice idea, but both halves of the equation are woefully inadequate: the sequence differences and the timescale calibration. First, protein sequence differences may not always provide a good measurement of relatedness. Second, the rate of species-wide mutation is so incredibly slow – millions of years per amino acid – that it must be estimated indirectly from the ancient fossil record that we wish to verify, creating a circular argument. But at least it's a start and, as always in science, we do the best we can, continually refining the approximations. Let's look at some nominal figures to get the general idea.

Imagine two presently living species thought to have a common ancestor 10 million years ago, according to the fossil record. A protein they have in common is compared, such as haemoglobin, a favourite because so much is already known about it and large databases can be consulted (to be precise, globin is the protein part and haem is a non-protein attachment). Their haemoglobins are found to be, say, 2 per cent different after 10 million years, providing the first point on the graph: haemoglobins might have diverged at a rate of around 1 per cent per 5 million years.

Repeat the experiment with pairs of species that are increasingly diverged. You may discover that species believed to have diverged about 25 million years ago are 5 per cent different, and 50 million years of divergence have led to 10 per cent difference. That, more or less, is what actually is found. Haemoglobins do seem to have diverged (changed their amino acid composition because of mutations in the genes coding for them) at a rate of about 1 per cent every 5 or 6 million years.

So what's the problem? Hasn't the dating problem been solved: just multiply the percentage difference by 5 and that's how many millions of years ago the two species had a common ancestor? Again, no! The problem is concealed in the weasel words: 'believed to have diverged', 'more or less', 'about 1 per cent'. The period per 1 per cent divergence turns out not to be a nice, consistent 5 million years but might fluctuate between about 3 million and 10 million years in different studies. That's just for haemoglobins. Different proteins have different divergence rates: a fast 1 per cent per 1 million years in fibrinopeptides, a slower 1 per cent per 20 million years in cytochromes, an exceedingly slow 1 per cent per 600 million years in one of the histone bobbins on which DNA is stored in the cell nucleus. These rate differences are understandable: for instance, changes to histones could fatally disrupt the storage of DNA and are almost never survivable, therefore changes are rare. Nevertheless, if different proteins can have such different mutation rates, how constant is the rate within any one protein?

Also, how do we know the crucial divergence dates when saying that species were thought to have diverged 20 million years ago? The best evidence is the fossil evidence – which is what we are trying to calibrate. Already this has become a tail-chasing exercise of trying to calibrate the rate of protein change from fossil evidence about the date of the ancestor, while trying to date the ancestor from the rate of protein change. It is easy to pick the wrong fossil ancestor, and in any case fossils are only dated indirectly from the estimated ages of nearby rocks. Worse, the older the fossil record, the greater the uncertainty.

Many species still living, such as small hard-shelled invertebrates, are practically identical to ancient fossils. The fossil crustacean *Colymbosathon ecplecticos* recently discovered in Britain, 425 million years old, has internal anatomy (revealed by the serial sectioning technique mentioned a few pages earlier) that is extremely similar to modern cousins. Has a species alive today descended, in the evolutionary sense, from a similar fossil species? The fact that the species has changed little in outward appearance in hundreds of millions of years makes it equally possible that it existed for hundreds of millions of years before the fossil was formed. Instead of the modern species being descended from the fossil species, both may date from a far earlier common ancestor. Even if the modern species is derived from the fossil species, did it diverge from the end, the middle or the beginning of the hundreds of millions of years when the species thrived? Which date should be used to determine the mutation rate?

The reality is that dating ancestors from fossil evidence is shaky, particularly as the correct ancestor is hardly likely to have been identified and dated with certainty. The measurement of relatedness from protein differences can be equally shaky, but to be fair, it is capable of replacing the uncertainty of anatomical comparisons with the certainty of exact measurements. If two species have the same DNA, they are the same species (loosely speaking: not even two humans, except identical twins, have exactly the same DNA, hence the significance of forensic testing). If they have just one different amino acid in a particular protein, there is no doubt that they had a common ancestor quite recently, meaning millions rather than hundreds of millions of years ago.

If a pair of proteins from different species differ little and their common ancestor was relatively recent, for example early humans or near relatives, then the measurement may be fairly reliable. Imagine how a population of animals or plants might diverge. A pregnant animal or a plant seed is carried to another island or across the mountains to start a separate population (even partner preference can be sufficient to cause a population to diverge, without geographical separation). Individuals in both populations continue to accumulate occasional germ-cell mutations that are passed on to their offspring.

As a reminder, mutations happen in the DNA of genes but the effect of a mutation is felt in the protein as a changed amino acid (or sometimes a deletion, insertion or something worse). One sometimes refers colloquially to mutant proteins, but the mutation event happened to the gene. Because every amino acid in the protein is coded for by a set of three DNA bases, each one of which can mutate among the alternatives C, A, G and T, it is exceedingly unlikely the same mutation will occur at the same time in individuals from the two separated populations. The different populations will accumulate different mutations.

Many of the theoretically possible mutations would be lethal and not be perpetuated through offspring. Others would not be especially disadvantageous or advantageous, just more or less neutral. This is important because a branch of evolutionary theory known as the neutral theory, championed by Motoo Kimura and others, predicts that a neutral mutation in one individual can spread throughout an entire population over a period of many generations. Mutations are rarely completely neutral and the current interpretation of the theory has been dubbed 'nearly neutral'.

Putting these various scenarios together – the isolation of two populations through geography or behaviour, different mutations appearing in each population, and some gene variants dying out – after a few million years the two populations will have distinctly different genes. DNA sequences will be a few per cent different, not only in haemoglobin genes but in all genes, and if the two populations can no longer interbreed they are regarded as different species. Evolution has happened and new species have been created. Over time the number of mutations increases further, protein sequences diverge and the two species become increasingly different.

Occasionally what appear to have become distinct species do in fact hybridise and produce fertile offspring. This is another mechanism of speciation.

In such an ultra-simplified scenario confined to the relatively recent past (but where there is sufficient divergence to have a reasonable number of mutations), sequence differences might be fairly indicative. Proteins diverge continually (which does not necessarily mean at a uniform rate) and counting the differences provides a way of measuring relatedness.

SETTING THE CLOCK

The principle of equating protein or DNA differences with time is known as a molecular clock. Provided that differences accumulate regularly over time, the idea is sound. Unfortunately, that's only the beginning of the problem as there are numerous sources of doubt. Here are three to be going on with. First, does 'regularly' mean a constant rate per year, or per generation of the life-cycle? Second, and either way, has the clock always ticked at the same rate in the past? And finally, are all clocks the same, for different genes or proteins and in different species? In other words, has a universal dating system been established?

To take the first problem, what does 'regularly' mean? Mice have far shorter lives than whales or humans and it would make all the difference whether mutations accumulated per year or per generation. While both points of view have support, there is more for the idea that absolute time counts in preference to the number of generations. This might seem counter-intuitive, but is logical if mutations are caused by natural radiation or chemical damage rather than something to do with cell division and fertilisation. A clue is to be found by studying 'dead' genes and long lengths of DNA that exist in the cell but are not apparently used for any purpose. As might be expected, the mutation rate is highest in this non-translated DNA because there are fewer negative consequences. This rate is probably near the maximum, that attributable to physical and chemical damage.

The second question is: has the molecular clock for a protein such as haemoglobin ticked uniformly over evolutionary time? Many scientists believe it has, but it's hard to prove definitively without specimens of very ancient DNA that are practically impossible to obtain. The consensus is that molecular clocks are a good guide within reasonable limits, and if the minute hand is wrong, at least the hour is about right. The further back in time one tries to project this idea of constancy, the more theoretically tricky it becomes.

The third question is whether all gene and protein clocks tick at the same rate. If they did, cross-checking with different proteins would be easy and we would have a superbly accurate clock for telling evolutionary time. But we have already said that they don't. Different types of proteins accumulate amino acid changes (through their genes) at quite different rates: histones extremely slowly, cytochromes much faster, haemoglobins faster still. Each has its own private clock. Taken to its logical conclusion, a truly constant evolutionary molecular clock is a self-negating concept, because the essence of evolution is change. If cytochromes, haemoglobins and myoglobins had a common ancestor, then before divergence the single ancestor obviously ticked at a single rate, yet now the three are different proteins with quite different rates. A protein sustaining successive mutations over hundreds of millions of years, strictly speaking, is continually becoming a different protein, so its mutation rate could be expected to vary.

The overall question was whether our relatives and ancestors can be dated millions of years back. The answer is yes – up to a point. Individual clocks may have ticked reasonably constantly since as far back as they can be traced. The trick is to understand the limits. Relatively modest protein divergences of less than about 20 per cent, or fossil ages of less than 100 million years, might give a reasonably good guide to the mutation rate. Although that's a very long time, it is barely the surface layer of the evolutionary pond, a mere 2 per cent of the age of the Earth. The other 98 per cent is very muddy and badly stirred up. We will never be able to see the bottom, but at least we can try to feel for it.

How can we feel through the murk? Feeling is not seeing, as we are reminded by the Indian fable of the six blind men who felt different parts of an elephant and had completely different ideas of what it was. What's needed is the whole picture. If the six blind men exchanged their experiences, they might come up with a whole picture of the animal that is greater than the sum of its parts.

The idea in the next chapter is to extract snippets of information that cannot be seen, from events that happened a billion years ago.

BEFORE THE ANCESTORS

Life is at least as old as the Earth

Relationship, as we have seen, can be measured as differences in protein sequence (or DNA sequence) between individuals, or species, because differences stem from genetic mutations that accumulate over time. By this means we can trace ancestry back millions of years, tens of millions, maybe hundreds of millions. That may seem a long time, but life is so incredibly ancient that a few hundred million years don't get us very near to the origin of life. The problem with going much further back is that mutations are overwritten on previous changes, obscuring history. But the grimy picture can be cleaned, and as with an art restorer peeling away centuries of dirt, varnish and Victorian fig-leaves, the revelations can be startling. Our cleaning solvent is the science of statistics.

HITTING AND MISSING

I hang a dart-board on the garage wall and start lobbing darts at it. No more random process is known to statistical science than my ability to aim darts at a dart-board, but just to complete the illusion I am blindfold and the dart-board is spinning (neither of which would make any difference). Since most of my darts would miss the board altogether (even if I cheated with the blindfold), only darts hitting the board are counted.

The board has the usual 20 sectors. The first dart to hit obviously lands in an empty sector, but that's crucial, so keep it in mind. As I continue throwing darts they arrive randomly in different sectors. Just occasionally a dart hits a sector that has been hit before. That's also very important. After 20 darts have hit we might find, say, two sectors with three darts in them, and three sectors with two darts in them, leaving eight sectors with one dart. That adds up to a third important observation, which is that seven sectors are still empty.

Having lobbed 20 darts into a dart-board with 20 sectors, obviously my average score is one dart per sector. But there's all the difference in the world between getting an *average* of one dart per sector, and *actually* getting one dart in each sector. In reality the distribution will almost certainly be nothing like one dart in each sector, as some sectors will collect two, three or occasionally more darts, leaving other sectors empty. The reason why the distribution will usually be uneven arises from one blindingly obvious piece of common sense, which is that otherwise, after throwing 19 darts, there would be absolute certainty where the 20th dart would hit – which is ridiculous since it can land anywhere.

By spending all weekend in the garage one could build up enough results to become very authoritative on the distribution of 20 darts among 20 sectors. The short answer is that with an average of one dart per sector the expected distribution is 37 per cent empty sectors and 63 per cent hit sectors, some of which must have more than one hit. More importantly, one could then predict with considerable accuracy what a typical result would be – on average.

Fortunately, there isn't any need to spend a weekend in the garage. Statistical science is able to arrive at the same prediction from first principles. As the number of darts on the board increases, the probability of hitting an empty sector reduces. Recall the mental note about the first dart being certain to hit an empty sector. Conversely, the last dart is confronted with more full than empty sectors, so is more likely to hit a full sector.

There are two clear extremes, or limits, to this game: all or none. Throw a large number of darts, say 100, and the probability of hitting an empty sector will be all (100 per cent) for the first dart, and none (virtually zero) for the 100th dart. Between the first and last darts the probability of hitting an empty sector will reduce gradually from 100 per cent to 0 per cent. Statisticians understand this very well and it is described by a surprisingly simple formula known as the Poisson distribution (shown in the Appendix).

Three important principles are learned from this kind of study, either by playing in the garage or by worshipping statisticians. First, the distribution of hits is uneven. Second, the general shape of the distribution – the chance of getting different numbers of darts in different sectors – is predictable from the Poisson curve (Figure 4.1). Third, there is no way of predicting which actual sectors will score multiple hits and which ones will be empty.

The Poisson distribution is still subject to the law of averages. While individual experiments – sets of 20 darts – will not necessarily give results exactly in accordance with the Poisson formula, otherwise once again the last dart would be compelled to land in the right place to satisfy the formula, on average they do. In statistician-speak, even the results of individual experiments are scattered around a mean. The more times the experiment is repeated and all the scores averaged, the more closely the average should resemble the perfect Poisson prediction.

Understanding how darts hit a dart-board helps enormously with unravelling mutations for the dating of life. The previous chapter introduced the idea that dividing the number of mutations by the mutation rate gives the number of years since a pair of species diverged. (Reminder: it's DNA that actually mutates, but this causes the protein to change and we usually compare the

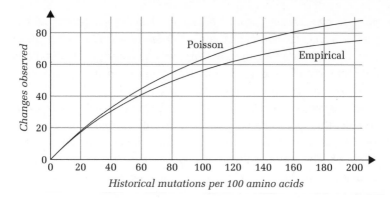

Figure 4.1 The Poisson curve corrects the observed number of mutational changes in a protein to the estimated historical number. The vertical scale is the number of differences between two sequences compared, *before* correction. The horizontal scale is the historical number *after* correction. The historical number has been compensated for successive mutations at the same site, which obscure previous mutations. Two different correction curves are shown. The upper curve labelled 'Poisson' is calculated accurately from the Poisson formula, but the correction is inadequate except when the correction is small. The lower curve labelled 'empirical', giving a larger correction, takes account of additional sources of error, but is less certain. From the work of Margaret Dayhoff.

protein sequences.) However, there are many reasons why the extent of divergence may be seriously underestimated.

Mutations in genes are like darts hitting a dart-board. They can strike anywhere. Individual bases in a gene can become mutated and replaced by a different base, but it is never possible to say with certainty that a position has mutated only once in a period of time. Just as a single sector on the dart-board can collect one dart after another, each base in a gene can undergo successive mutations. The base may change again and again, and so the amino acid in the protein may change again and again, leaving no record of the previous changes (although there are cunning ways of finding clues that this has happened). The apparent percentage of mutations therefore always underestimates true mutational history, because mutations overwritten at the same site cannot be seen.

This is where the Poisson formula becomes useful, because it corrects this underestimate to some extent. Suppose that proteins are compared from two species and the sequences are 20 per cent different. The graph in Figure 4.1 shows that 20 per cent observed changes can be corrected to 23 per cent actual historical mutations. Assuming proteins 100 amino acids long, they seem to have 20 differences, but historically there have been 23 changes, 3 of which must have changed an amino acid that had changed previously.

At least, that's the theory. It works well for modest numbers of changes, but unfortunately not well for large differences and to get a better correction we have to dig a little deeper. Checking the Poisson curve (upper curve, Figure 4.1), three phases can be made out. If the observed difference between two sequences (vertical scale) is small, say 10 per cent or 20 per cent, the correction (horizontal scale) on the Poisson curve is minor. If the difference is moderately large, say 50 per cent, the error is large and the correction is substantial. And if the difference is very large, such as 80 per cent, the correction is unreliable to the point of being virtually meaningless. The reasons are important.

SMALL CORRECTIONS

The first phase of the curve is where there are few mutations. The two species being compared diverged a few millions or tens of millions of years ago. As a pair of sequences start to diverge, the first mutation is certain to be at a fresh site, one that has not mutated before. The second mutation will strike at random and would require an extraordinary coincidence to hit the same site, so is almost certain to hit a different site. Because further mutations accumulate at random, there's nothing to stop a mutation from hitting a site that has been hit before, but the chance of this happening is small.

By the time 10 per cent of sites have mutated, the next mutation has a 9 out of 10 chance of hitting a fresh site but a 1 in 10 chance of hitting a site that has mutated once before, changing it a second time. The Poisson correction for multiple hits is still quite trivial and even at 20 per cent sequence difference the Poisson correction contributes only another 3 per cent. Such corrections are small in comparison with all the other uncertainties involved, and therefore fairly reliable.

LARGER CORRECTIONS

In the second phase of the curve, above about 30 per cent mutations, the problem compounds rapidly. Once 50 per cent of sites have been mutated, the next mutation is just as likely to hit a fresh site as an old one – or possibly a site that has mutated several times already. A measurable difference of 50 per cent means that the predicted number of historical events is really 70 per cent. That's a substantial error, and as the number of mutations rises still further the problem escalates towards the third phase, where virtually all sites have mutated at least once. Then it becomes a near certainty that every new mutation will change what has already been changed at least once before.

LOSING TRACK

At the top end of the Poisson curve the ability to make a sensible correction fizzles out altogether. The flattening of the curve at the top, where observed amino acid changes exceed 80 per cent, says that to change most of 100 sites would require a surprisingly large number of actual hits. The number is one of those vanishing targets that is never quite reached: 90 observed changes result from 230 hits, 95 changes result from 300 hits, 99 changes result from 460 hits, and so on (using the formula – these are off the end of the graph). The method no longer works.

And worse is to come. The Poisson curve, while a clever way of seeing into the past, is itself inadequate, particularly for the distant past. This is because mutational events can be more complicated than the pure dart-board model. For instance, two amino acids that seem identical may both have changed the same way, some DNA changes do not change the amino acid, and there are other reasons.

Various attempts have been made to bundle all these special extra considerations, together with the Poisson correction, into a single correction. The lower curve drawn on the graph, labelled empirical, incorporates these additional corrections. The empirical curve is so much flatter than the pure Poisson curve at the high end that a measurable difference of 75 per cent converts to about 200 changes per 100 sites, and an 80 per cent observed difference leaps to become roughly 300 per cent historical changes. That's equivalent to a pair of sequences changing three times at every amino acid, or one and a half times each, on average. Clearly, a practical reliable limit exists above which protein sequence comparisons blur into obscurity. Taking into account the empirical corrections, it looks as though a 75 per cent difference is about the limit.

The difference between the two correction curves is that the Poisson curve, being calculated from a mathematical formula, is exact but not sufficiently correct at the higher levels, whereas the empirical curve is more correct but less exact.

As it happens, there are independent reasons why a pair of proteins (still doing the same job) will not diverge much more than about 80 per cent in any case. To begin with, even random proteins would never be completely different. Look along the lines of this page and you will find occasional coincidences where a letter on one line happens to be the same as the one above (I have not even checked because it is bound to be so!). In the same way, a pair of random protein sequences printed one above the other would have the same amino acid at odd points by chance. Since there are 20 amino acids to choose from, each amino acid in the top sequence would have a 1 in 20 chance of aligning with the same below – in other words, 5 per cent agreement would be expected.

Then there's the fact that functioning protein molecules are not merely a linear sequence of amino acids, they have three-dimensional structure. The sequence, once assembled, folds itself into the extremely precise shape of an enzyme, or haemoglobin, or silk, or whatever it is destined to be. The wrong amino acid in a critical place could be as devastating as a car engine having a wrong-sized piston, so in these places divergence is not allowed. In some places a protein might allow a changed amino acid, as mentioned in Chapter 3 (that's how evolution occurs), but in other places the choice might be extremely restricted. Haemoglobin, our oxygen transporter, is never found with its key histidine amino acid (number F8) substituted with anything else, otherwise it could not properly hold the essential iron atom that carries the oxygen.

So for two reasons, unravelling evolutionary divergence by means of protein sequences fades out at about 75 per cent or 80 per cent divergence. One reason is that the empirical curve, although the more correct, is really thick and fuzzy at its top end and cannot be read accurately. The other is that comparable proteins do not attain sequence differences of more than around 80 per cent anyway.

The effect of uncertainty at the top end of the curve is that divergence history is far more likely to be underestimated than overestimated. If the conversion from observation into history fades out at around 80 per cent observed changes, then the converse is crucial: however long the history may be (taken to its logical conclusion, infinitely long), observed divergence would not exceed 80 per cent. History, and therefore age, becomes imponderable.

Haemoglobin is a model protein extensively studied in this way. We would like to know the ancient history of the molecule because it is of great importance as the carrier of oxygen in our blood. The prospects for discovery are good, as it has widespread occurrence in different organisms and the database of studies is large, making it easy to cross-check findings for consistency. The generally accepted calibration for haemoglobin is 1 per cent divergence per 5 to 6 million years. Provided that this rate has remained reasonably constant for a very long time, then taking the empirical curve, a 75 per cent difference converts to 200 changes per 100 sites and a timespan of $5 \times 200 = 1000$ million years (1 billion years). Comparisons between haemoglobins often reveal even higher differences of 80 per cent or more, corresponding to 300 (or possibly far more) changes per 100 sites. Such differences correspond to at least 1.5 billion years but are well into the hopelessly unfathomable region of the curve and could be older still. We have hit a limit that corresponds to at least a billion years, some would say 1.5 billion, or even older.

But that's the point: the method has a natural limit and loses precision when changes mount upon changes and history becomes a blur. The figures are the best we have and simply show that haemoglobin is very, very ancient. As far back as can be seen with sequence comparisons, haemoglobin existed. To place 1.5 billion years in perspective, it is more than twice the age of multicellular organisms and more than three times the age of land animals. Much-studied haemoglobin is a surrogate for numerous other complex and highly developed proteins and the overall conclusion is that many, if not most, are exceedingly ancient.

SPLIT GENES

Much has been said about mutations, implying that a mutation is the simple replacement of one of the bases (C, A, G or T) in a gene with a different base. Consequently a different amino acid may be incorporated into a protein (or it may not – check the translation code from gene to amino acid in the Appendix). As mentioned previously, the replacement amino acid may be acceptable to the protein but on the other hand it may be unacceptable, crippling its function.

Mutations can be more devastating than simple replacements. If a base is deleted altogether from DNA, or an extra base is inserted, the neat translation of bases three at a time into amino acids is thrown completely off the rails, just as omitting a digit when dialling a phone number, or slipping in an extra digit, will not get you even close to the person you want.

Still greater upheavals can cause large chunks of gene to be moved around. The complete set of genes (the genome) for, say, a human is not arranged as a continuous text read from start to finish. Oddly enough, most of it is unreadable nonsense. The relatively small parts that are meaningful are themselves interrupted with seemingly meaningless tracts.

Halfway through a page 1 newspaper article, in the middle of a sentence, there's a hyphen and it says abruptly 'turn to page 29, column 7'. Infuriatingly, page 29 has been used to light the barbecue. Many genes are arranged like that, with chemical signals in their midst that suddenly say 'jump several pages and then pick up the rest of the story'.

The useful information in a gene might be a few hundred bases in length. Every amino acid in a protein is coded for by a codon of three bases, so a haemoglobin-sized protein of 150 amino acids is encoded by 150 codons, or 450 bases of DNA (Figure 4.2). But the gene itself may be very much longer. In the 1970s it was discovered quite unexpectedly that many genes are interrupted with long tracts of non-translated sequence. The interruption can start between codons or even in the midst of a codon. The interruption can be many times larger than the informative parts of the gene and a single gene can have several such interruptions.

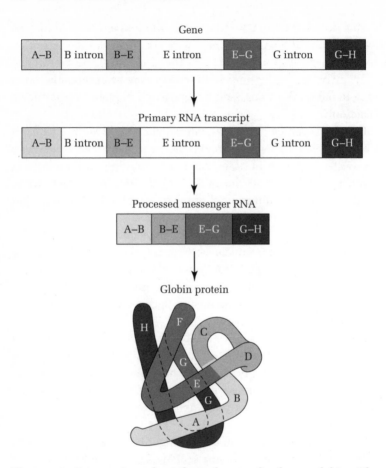

Figure 4.2 Introns in a hypothetical gene for haemoglobin. The haemoglobin protein, bottom diagram, is a continuous sequence of amino acids folded into a unique shape. Its main regions are known by the letters A–H. The DNA, top, contains the same information but has three inserts labelled as introns. The information content of the DNA is first replicated as a full-length RNA transcript, which the cell then edits, saving only the information needed to make the haemoglobin.

These lengths of intervening sequence are called introns to distinguish them from the informative parts of the gene, called exons (intron for intervening, exon for expressed). An early term for the phenomenon was 'split genes', a phraseology inclined inexplicably to provoke student mirth. Somehow the cell's machinery has to be instructed to read, say, bases 1 to 149, then jump and read 1900 to 2050, then 4851 to 5000, a total of 450 bases, and use that information to make an uninterrupted protein sequence of 150 amino acids. Base numbers 150 to 1899 and from 2051 to 4850 are to be ignored.

The actual cutting-out process is not done to the DNA or to the protein, but to an intermediate copy of the information known as RNA (Figure 4.2). RNA is a nucleic acid fairly similar to DNA but single stranded (and has U, uracil, where DNA has T, thymine). The information content (DNA sequence) of the gene complete with introns is first replicated as RNA, then the introns are cut out, and finally the edited RNA, containing only the protein information, directs the assembly of the protein sequence.

Introns are located in genes but are sometimes loosely referred to by their equivalent position in the protein coded for by the gene, because this is more illustrative. Another essential piece of terminology is that talk of tracing the evolutionary history of an intron refers to the position of the intron in the gene, not the intron's own internal sequence, since introns themselves mutate rapidly, including enormous changes in length.

Introns are important, for three reasons. First, they appear to be very ancient and provide a window into an earlier period of life than can be seen by protein sequence comparisons. Second, such a seemingly unnecessary disruption of genetic orderliness may reveal secrets of how very early life developed. And third, introns and their retinue of associated enzymes for processing them are further evidence that early life was highly complex. As with many components of life, it is not immediately obvious how introns and their editing systems could have evolved gradually, but equally unclear how they could have appeared suddenly.

The two main theories of where introns came from are known as *introns-early* and, unsurprisingly, *introns-late*. Briefly, *introns-early* proposes that introns had a primordial origin. The theory proposes that large, complex proteins were originally composed from smaller, compact mini-proteins, effectively modules, maybe much earlier than a billion years ago. Small genes expressing small proteins were joined together to form large genes expressing large proteins. The key part of the *introns-early* theory is that the small genes did not become joined exactly end to end because they already had a certain amount of non-translated flanking sequence, like the blank ends of a roll of film. The joined-together sequence therefore consisted of segments of meaningful DNA ready for translation into protein, alternating with segments of non-translated DNA.

This is a handy way for complex proteins to evolve, drawing on a kitset of small modules, but the flip side is that a gene composed of alternating meaningful and non-meaningful lengths of DNA presents the cell with real problems. The cell has to process a length of meaningful genetic information, then jump ahead quite a lot, then process another length and so on, joining it all together. As mentioned earlier, the cell achieves this by editing the RNA transcript of the gene to remove the unwanted bits. The corrected messenger RNA is then translated to assemble the corresponding protein sequence.

Knowing how this is now achieved comes nowhere near to explaining how such a complex mechanism could have evolved. The editing mechanism is convincingly close to a display of intelligence, although as we continually remind ourselves, it is only chemistry. Somehow the cell's machinery has to 'know' which segments of RNA to cut out and then has to join the good bits together. This has to be a matter of recognising the codes of chemical structure in the RNA sequence. The really difficult concept is how this could have evolved gradually, but the same conundrum of apparently irreducible complexity applies to all of life.

The alternative *introns-late* hypothesis is quite different. The general idea is that introns were inserted into perfectly good genes at a later date. This could still have happened aeons ago, which is why references to early or late can be misleading and better words would be 'formative' and 'insertional'. The late or insertional theory proposes that the gene for a protein such as haemoglobin was originally uninterrupted and free from introns. Presumably the gene had evolved gradually, changing its sequence and growing in length as the protein it expressed evolved in complexity. That is itself a difficult concept. Then, suddenly, an intron got inserted into the gene, and others later; still more puzzlingly, the cell somehow had access to the wherewithal to edit them out of protein synthesis.

A possible hybrid theory is that at least some introns originated with the joining of mini-genes (*introns-early*) but that most of them have since disappeared from genes because they never really served any purpose. Some of the surviving introns have been moved around or inserted elsewhere by the insertional mechanism. A more recent theory has even smaller and more primordial modules ('trexons', transposable exons) shuffling themselves around.

Introns give a tantalising glimpse of life earlier than can be traced by sequences, because their locations appear to be more stable. It is the location of the intron (not its internal sequence) in the gene that is stable, but that location can be defined only by marking it on the protein. For example, haemoglobin normally has two or three introns, one of which is almost invariably in the part of the gene that, when translated, codes for the 12th amino acid in the B helix of haemoglobin structure (Figure 4.2). That intron location is strongly conserved in globin genes of families as widely diverged as plants, vertebrates and invertebrates, more conserved than the gene or amino acid sequences themselves (which may be

80 per cent different). At the very least we can see that complex proteins existed long before the billion-year-or-so scope of protein sequencing. We are now peering into the 2-billion-year era, but can travel back further yet, before returning to introns in Chapter 9.

REALLY ANCIENT PROTEINS

When the first atomic bomb was being concocted in the 1940s, it was realised that it would need to be built from almost pure uranium 235 (^{235}U). Natural uranium contains only a 0.715 per cent trace of the isotope ^{235}U, the more than 99 per cent remainder being mainly ^{238}U. Unfortunately – or fortunately – ^{235}U and ^{238}U are both chemically uranium and it is impossible to separate uranium from uranium by chemical means. The only answer was to separate the atoms physically, by size.

The method chosen was gaseous diffusion, which depends on ^{235}U being very slightly lighter than ^{238}U. Uranium was converted into uranium hexafluoride gas and diffused through porous material, exploiting the slightly faster diffusion of the lighter ^{235}U hexafluoride. Since ^{235}U hexafluoride is lethally radioactive, corrosive, highly toxic and a gas able to escape through minute flaws, and the gaseous diffusion process requires thousands of repetitive stages, the technology to build a safe and reliable separation factory is prodigious and mercifully beyond the capability of most nations. (The alternative atomic bomb uses plutonium, which can be separated from uranium chemically, but first one needs to build a nuclear reactor to generate the plutonium.)

Living organisms have been separating isotopes by a process similar to gaseous diffusion for years. Billions of years. As a result, we can identify traces of life from billions of years ago. One of the elements that life has been sorting isotopically is carbon. All life on Earth is carbon based. Proteins, enzymes, nucleic acids, fats, carbohydrates, cell walls, are all carbon compounds. The primary source of all this carbon is the mineral reserve, through the massive bicarbonate content of the oceans, and the much smaller reservoir of atmospheric carbon dioxide. Dissolved carbon dioxide is more or less the same as bicarbonate.

The point at which the majority of carbon enters the organic chain is when plants take carbon dioxide from the air or bicarbonate from water and incorporate it into simple sugar-like compounds. These compounds are then assembled into the biochemical components of the plant, such as cellulose cell walls. Animals obtain most of their carbon by eating plants (or other animals that have eaten plants further up the food chain) so ultimately all life is built from plant carbon, except for a small amount of carbon dioxide exchanged with the atmosphere by animals.

Carbon in the ocean reserves is a mixture of nearly 99 per cent of the isotope carbon 12 (^{12}C) and just over 1 per cent carbon 13 (^{13}C). Carbon in the organic molecules of plants and animals is invariably found to be very slightly deficient in ^{13}C compared to inorganic mineral carbon. The difference is quite small, organic carbon containing on average around 1.085 per cent ^{13}C compared with approximately 1.11 per cent ^{13}C in ocean bicarbonate, a difference of 25 thousandths of 1 per cent. Nevertheless, this is a priceless discovery that enables traces of life in ancient rock to be identified because their carbon has the organic rather than the mineral ^{13}C signature. It doesn't date the rock, but that can be done independently.

The mineral carbon isotope ratio differs slightly in different places and so ^{13}C depletion is traditionally calculated against an arbitrary standard known as PDB, which is a story in itself. When this sort of work started, a sample of limestone dating from the Cretaceous period, collected from the PeeDee river formation in South Carolina, was adopted as a standard for comparison. It became known as the Chicago PDB Marine Carbonate Standard, the letter B meaning belemnite-type limestone. The limestone is mainly the fossil of the hard inorganic carbonate guard from the pointed end of the extinct squid or cuttlefish *Belemnitella americana*. Cuttlefish are not fish but molluscs, related to snails, and the so-called cuttlefish bone (analogous to the guard of *Belemnitella*) put into bird cages is not a bone at all, but the equivalent of a shell that has been internalised to give the squid some rigidity. Cuttlefish shells are assembled mainly from inorganic carbonate taken directly from the sea and research has established that their carbon isotope composition closely reflects the seawater, and thereby the mineral reserve. A tiny amount of truly organic carbon might also be present because cells regulate the formation of the shell.

The main thing is to have a reference standard, and by agreement PDB is it. Unfortunately the original standard has all been used up, but results can still be scaled back to PDB. A specimen with the same ^{13}C content as PDB would have a depletion of 0 parts per thousand, a sample devoid of all ^{13}C would have a depletion of 1000 parts per thousand. The ^{13}C depletion in biological samples is usually between 10 and 40 parts per thousand, or typically 25 parts per thousand in very round figures.

How does this difference come about? The incorporation of carbon dioxide into organic molecules is, like virtually all biochemical reactions, catalysed by enzymes. One of these enzymes, known as rubisco (short for ribulose bis-phosphate carboxylase-oxygenase), is so widespread in the plant world that it is effectively the point of entry of all carbon into the biosphere (Figure 5.1, next chapter). When carbon dioxide with its mixture of mainly ^{12}C and a little ^{13}C reacts with rubisco, there is a slight discrimination in favour of ^{12}C because, being less massive than ^{13}C, it diffuses slightly more rapidly into the accurately fitting reaction site on the enzyme. Rubisco is mainly responsible for the ^{13}C isotope deficiency characteristic of all biological carbon. An article by Manfred Schidlowski, who has contributed much to this study, is referenced in the Appendix.

Organic ^{13}C depletion is measurable in ancient rocks that were formed by the sedimentation of organic material, and the depletion is always in the same broad range, centred around 25 parts per thousand. Some rocks revealing this tell-tale sign of biological origin are extremely ancient indeed. Minik Rosing at the Geologisk Museum, Copenhagen, has found the tell-tale isotope signature of biological carbon in rocks from the Isua formation in West Greenland, generally agreed to be sedimentary and around 3.7–3.8 billion years old. It's important to be sure that this is not a circular argument and to check the independent lines of evidence.

First, there is the age of the rock. The ^{13}C to ^{12}C ratios do not date the rocks and must not be confused with radiocarbon (^{14}C) or other isotope dating. (Measurements of ^{14}C are useful for dating only very young organic remains, up to 60 000 years old, because ^{14}C practically all decays in that time. There would not be any on Earth except that it is created in the upper atmosphere, but after organisms die it is not replaced in their remains.) The very ancient rocks we are talking about are dated independently through the presence in nearby rocks of isotopes having nothing to do with carbon, such as uranium and lead (Chapter 1).

The question of whether rocks are of mineral or partly organic origin is determined independently by means of their chemical signatures and from indications of how the rock was formed, such as by sedimentation. Stephen Mojzsis has dated a specimen from Greenland's Akilia Island up to 3.87 billion years old, slightly earlier than the Isua deposit, but recent studies have undermined this finding with geological evidence that the deposit, although layered, may not in fact be sedimentary or organic.

Finally, the ^{13}C to ^{12}C ratio is another independent measurement. Putting together these results from many thousands of rock samples of various ages from all over the world, the conclusions are consistent: first, organic constituents of rocks are always deficient in ^{13}C compared with mineral rocks; and second, the deficiency has been broadly consistent since at least 3.5 billion years ago and probably for 3.8 billion years.

The size of the carbon mineral reserve in rocks and oceans is so great compared with the biosphere that even over the timescale of nearly 4 billion years, the isotope ratio of mineral reserves has not been significantly changed by the effects of life. Mineral and biological carbon are distinctly different from each other and the ^{13}C depletion identifies them. The atmosphere, however, is quite a small reserve and this has accumulated a depletion of about 7 parts per thousand through biological activity.

The conclusion is difficult to escape: life not only existed 3.8 billion years ago but was sufficiently prolific to leave behind sedimentary beds that formed rocks enduring to this day. But that's indirect evidence: where are the fossils, pictures and photographs that are more likely to convince people? In many ways the ^{13}C isotope record is so continuous and so consistent as to be even more convincing than certain debatable signs of cell-like microfossils in ancient rocks.

Over a period dating back to the early 1980s, William Schopf in the USA took photographs of microscopic inclusions seen in thin slices of rock from sites near Marble Bar, inland from Port Hedland, Western Australia. The special significance of the rock was that it was very ancient, dated to about 3.5 billion years ago, and appeared to be sedimentary and therefore a place to search for fossils. Schopf interpreted some of the microscopic shapes visible as being chains of fossilised micro-organisms broadly similar to modern cyanobacteria. The latter, sometimes inappropriately called blue-green algae, are cyan-coloured bacteria that photosynthesise similarly to plants. The fixation that many researchers have on early cyanobacteria is that photosynthesis, by releasing oxygen, is the likely source of all atmospheric oxygen, and cyanobacteria are a convincing model of an early photosynthetic organism. Schopf's photomicrographs, a couple in particular, have appeared in numerous textbooks as the oldest known fossils and evidence that early life on Earth was cellular and, by implication, may have been photosynthetic like cyanobacteria.

As so often happens after a piece of research achieves classic status, sooner or later someone decided to revisit the discovery. Martin Brasier in Britain, after examining Schopf's original specimens and visiting the site in Australia, came to the conclusion that the rock was not sedimentary in the accepted sense and that the inclusions were not fossilised cyanobacteria, or even cells, but structural artifacts. Similar artifacts have been recreated recently by Juan Manuel Garcia-Ruiz in Spain. The debate will doubtless develop as others join in and is far from resolved.

Meanwhile, where else can we look for signs of primordial life? Perhaps visible fossil evidence would be almost too good to be true and corroboration of the ^{13}C findings must depend on independent forms of evidence. One of the most convincing is the work of Jochen Brocks (mentioned again in the next chapter).

Brocks and colleagues, also working in Western Australia, analysed complex molecules present in shale drilled from a depth of 700 metres. Again, the shale was a good place to look because the oil it contains is considered to be of biological origin and this particular deposit is about 2.7 billion years old, not as old as the Isua deposits but pretty ancient nevertheless. The discoveries were fascinating; for instance, they found 2.7-billion-year-old molecules characteristic of cyanobacteria, and long chains characteristic of the membranes of eukaryotes ('higher' cells) rather than prokaryotes (bacteria). But the date of 2.7 billion years is between half a billion and a billion years older than the previously accepted antiquity of eukaryotes. The forms of life that left behind these molecular fossils were variously complex and photosynthetic, and, by logical extension, had already taken a considerable period of evolution to reach such states of complexity.

DNA studies by Carrine Blank at Missouri, using the sort of sequence comparisons mentioned in Chapter 3, say that cyanobacteria evolved about 2.4 billion years ago. This does not conflict with the earlier date of 2.7 billion years suggested by Jochen Brocks and the much earlier dates from ^{13}C analysis, because no one is claiming that cyanobacteria, as we know them now, are exactly the same as micro-organisms in existence that long ago. The deductions are simply the reverse, that signs of life 2.7 or even 3.8 billion years ago had characteristics now found in cyanobacteria – photosynthetic, membrane-bound microbes.

PRIMORDIAL LIFE: A COMPREHENSIVE PICTURE

The ^{13}C isotope signatures of life date back 3.8 billion years or so. Molecular fossils indicate that life not only existed 2.7 billion years ago but had achieved considerable complexity, including photosynthesis (as previously mentioned, plant cells performing both photosynthesis and respiration are more complex than animal cells, which only respire). Eukaryotic life had been established, whereby cells had advanced beyond the simple bacterial type by parcelling out internal functions to different organelles.

Even if judgement is reserved on visible microfossils, the multiple facets of life's ancient traces, interacting together, support a remarkably comprehensive vision of primordial life on Earth. Life was prolific, microscopic, cellular, membrane bound, and some at least was photosynthetic, biochemically based on the fixation of atmospheric carbon dioxide into organic chemicals, which invariably contain substantial hydrogen. These are profound deductions because, among other things, they imply the extraction of hydrogen from somewhere, the most abundant source being water, which inevitably means the release of oxygen. The oxygen on which nearly all life now depends was made by life itself. Life's not simple – and it's difficult to see how it ever was.

LIFE'S NOT SIMPLE

Life on Earth has always been complex

The direction of the argument so far is plain. Life 3.8 billion years ago was already exceedingly complex. To all intents, life at the biochemical level was as complex then as it is today. We have barely scratched the surface of the myriad chemical reactions necessary to assemble just one single protein, but the process is extremely well understood. No specific protein such as rubisco (Chapter 4) could have been constructed that long ago without a gigantic inventory of biochemical pathways to supply the amino acids, the energy sources such as ATP and GTP, the base-sugar-phosphate complexes (nucleotides) for making RNA, and numerous other essentials. Protein assembly cannot happen without the hugely complex intracellular machine known as the ribosome, itself constructed from numerous proteins and RNA. Many other indispensable pathways have not been mentioned, such as the synthesis of the haem component of cytochromes and haemoglobin, itself a daunting task. In short, there are very few components of modern cells that could have been left out billions of years ago, otherwise life would not have worked.

This does not deny that life was simple before it became complex, a matter that will be discussed more in Chapter 7. What it does deny is that life inside the cell, on Earth 3.8 billion years ago, was much simpler than it is now.

WAS LIFE EVER SIMPLE?

It would have been a tempting excuse to say that the biochemistry of cellular life a billion or more years ago must have been far simpler than now. But the idea simply won't fly. All modern forms of life have an extremely comparable set of machinery for copying and translating genetic DNA for the ultimate manufacture of proteins. Just as the similarity of modern proteins across the spectrum of life points to a common and ancient origin, so does the similarity of the machinery for their manufacture. Essentially, the same set of devices for translating DNA, via RNA, into protein was present in the primordial ancestors. This is obvious because the process is so similar in all modern forms of life. If this complexity had developed recently, different branches of life would have introduced fundamental variations into the synthesis mechanism. They haven't.

The directed synthesis of a complex protein could not have happened in isolation and required virtually all of the components of a modern cell. But those thousands of components are themselves either proteins or are produced by the action of proteins. Proteins include enzymes, structural proteins, membrane permeability proteins, hormones and many more. Products of proteins include nucleic acids (DNA, RNA) copied through the mediation of highly complex enzymes, also cellulose cell walls of plants, small molecules such as ATP and numerous others. The whole existence of cells is directed towards the manufacture of proteins that in turn are the manifestation of life. However colossal and wonderful the information content in the DNA of a human nucleus, that information has achieved nothing until it is materialised in the form of proteins.

Some of the questions previously posed are beginning to provide answers to each other. The difficulty in accounting for a protein as complex as haemoglobin as long ago as a billion years is nothing compared with accounting for rubisco 3.8 billion years ago. Yet on the principle that carbon dioxide-fixing life seems to have been complete in itself, a clear picture emerges of flourishing primordial life as being microscopic, photosynthetic and sufficiently prolific to leave behind large deposits of carbonaceous sedimentary rock.

PRIMORDIAL LIFE WAS EXTREMELY COMPLEX

If cellular life existed at least 3.8 billion years ago, how were the hundreds of proteins and essential components made that long ago? If the enzyme rubisco has indeed been the primary mechanism responsible for the fixation of biological carbon since not long after the Earth was able to support life, accompanied by numerous other complex enzymes, how were these manufactured in primordial times? The short answer has to be, the same way as they are now!

Take just the first step. The genetic DNA carrying the blueprint for rubisco or haemoglobin or any other protein is not fed directly into the protein synthesiser (the ribosome). Instead, the intelligence stored in the DNA is replicated in the form of messenger RNA (mRNA, Chapter 4) that carries the instructions to the protein synthesiser. The assembly of RNA itself requires the intervention of some of the most complex enzymes known to biochemical science. In this process, the DNA double helix is unwound locally by a special enzyme, while RNA is pieced together from exactly the right nucleotides in the right order to convey the correct message. But nucleotides, the building units for RNA, are themselves complex molecules containing the sugar ribose, phosphate groups, and a coding base (A, C, G or U; note the small differences from DNA (Chapter 3) which has A, C, G and T, and deoxyribose). These nucleotides are synthesised through biochemical pathways, each a succession of different reactions in which the molecules are constructed stepwise, every step synthesised by a dedicated enzyme. All these enzymes are also proteins, products of the same system.

The big picture is beginning to emerge. Not only is the conversion of genetic instructions into protein a highly involved process involving numerous steps, but each step is fed by an array of metabolic pathways to supply all the necessary ingredients.

The protein synthesiser is another example of incredible intricacy. The synthesiser is known as the ribosome, a large molecular structure inside the cell where amino acids are fitted together to make proteins. Each cell contains thousands of ribosomes. The functions of the ribosome are so intricate that it's possible when working with them to get carried away by the things they 'do', almost as though they are small creatures. The cold reality is that ribosomes do no more than catalyse chemistry when molecules collide spontaneously.

With the help of the ribosome a succession of amino acids are held together in the correct order to assemble the protein, stepwise. The information for that correct order is the order of bases on the messenger RNA, three bases (a codon) coding for one amino acid. The ribosome mediates the chemistry of matching together each set of three bases to one amino acid (through the adaptor molecule transfer RNA, tRNA), followed by the chemical joining of the amino acids to assemble the protein. The reason that antibiotic drugs kill bacterial infections so effectively is that they block protein biosynthesis in bacteria at just one of these numerous stages, depending on the antibiotic, which indeed is how much of the detail of protein biosynthesis was discovered. (Many known antibiotics discriminate the opposite way and would kill the patient while the infection survived, but there's not much call for them.)

As can be imagined, the ribosome is a complicated piece of machinery. The simplest ribosomes as found in bacteria contain no fewer than 50 different protein molecules (each, at the risk of becoming repetitive, a product of the same system) together with some important lengths of RNA. Ribosomes in higher organisms are larger and more complex, containing about 80 protein molecules, but they do the same job – another example of biological structures being either similar or different depending on one's point of view. As before, each of the reactions catalysed by the ribosome requires a stream of extra ingredients such as GTP (an energy source rather like ATP) plus the supply of amino acids, all 20 types, all needing to be made through metabolic pathways by means of a battery of dedicated enzymes (again, each a product of the same system!). The purpose of this detail is to press home a recurring and inescapable point: primordial life on Earth was astonishingly complex. It cannot have been significantly simpler then than it is now.

WRAPPING UP LIFE

Yet the story is nowhere near finished. The biochemistry of life cannot proceed unless the components are present at the relatively high concentrations found inside cells. If a cell is punctured and the contents mix with a little water in a test-tube, the cell is dead. The components are still there and a few of the reactions might proceed, slowly, but the integrated functions of the cell amounting to life would not happen. Life is no longer recognisable. That is precisely why cells exist. One key to life is that the very small volume of the cell keeps the enzymes and biochemicals at a high enough concentration to allow the reactions to proceed at a realistic rate. Another is that many reactions are driven by concentration differentials existing either side of a membrane. Early life happened in cells, and cells are complex. They are enclosed within membranes, which themselves are complex.

A membrane is not like an impenetrable plastic bag, otherwise nothing could get in or out. It has to be sufficiently permeable for fuels and nutrients to enter, for unwanted waste products to leave, and to exchange carbon dioxide and oxygen with the air. On the other hand, a membrane permeable to every type of molecule would be useless, as it might as well not exist. What's needed is a membrane permeable to some things and impermeable to others, in other words selectively permeable.

Selective permeability is another illustration of the inescapable complexity of life. Membranes achieve their selective permeability by means of a partnership between fatty lipids that provide the impermeability and proteins that provide the permeability. The lipid makes the membrane impermeable to water and most dissolved chemicals. Some small molecules such as gases and also lipid-soluble chemicals (such as steroid hormones) can diffuse through the lipid part of membranes unaided, but larger molecules are only allowed through by special proteins, resembling little trapdoors embedded in the membrane. Many of these trapdoor proteins are highly sophisticated and allow an incoming molecule to exchange for a different outgoing molecule, or open to allow a particular molecule to pass through only when it benefits the cell. Different tissues and organs have cells that differ in the kinds of molecules they let in or out, which is partly why the various organs function differently.

The so-called problem of irreducible complexity immediately springs to mind. How could primitive cells have existed without their modern array of special membrane proteins that so exquisitely regulate communication with the outside world? How could they have existed with simpler, less evolved membranes that were either much more permeable or much less? How could a cell possibly respond to the need to (say) import sodium and export potassium by developing special membrane proteins to let them in and out? How could the cell exist before it developed the solution to the problem, and how could the problem be solved before the cell existed?

Questions of this sort are formulated back to front. The real way to look at it is that more primitive cells would indeed have had simpler membranes, with a rather different spectrum of permeability, but would have been far less efficient. The efficiency of life increased as primitive organisms refined their membrane permeabilities, acquiring (by evolution and selection) the sophisticated permeability proteins now characteristic of different types of cells, depending on their job. Cells function with the set of selective permeabilities that have become available and have been retained; they cannot seek the impossible.

Overlaying the protein synthesis system is another whole layer of regulation to ensure that each type of protein is made only when required. It's no use a cell simply churning out its full repertoire of proteins the whole time. The cell would fill to capacity with all sorts of proteins, most of them unnecessary and some absolutely dangerous. Life's not like that. Different proteins are made in the precise amounts required to do specific jobs at the right times. Specialised cells produce insulin, growth hormone, blood albumin, antibodies, hair, muscle fibres, the collagen that gives strength to skin, a huge catalogue of dedicated proteins. Such production is extremely finely regulated to ensure that the correct protein is available in the correct amounts in the right place at the right time. A single bout of overproduction or underproduction of a protein as critical as insulin could quickly be fatal. There would be no second chance.

ENERGY AND OXYGEN

The oxygen that most life depends on was put into the atmosphere by life. Science is remarkably close to unanimous on that wonderful piece of circularity.

Plant life releases oxygen. The basic biochemistry of plant life is to construct carbohydrates, and thereby its whole self, by combining carbon dioxide with hydrogen. The hydrogen is obtained by splitting water, H_2O, which unavoidably leaves

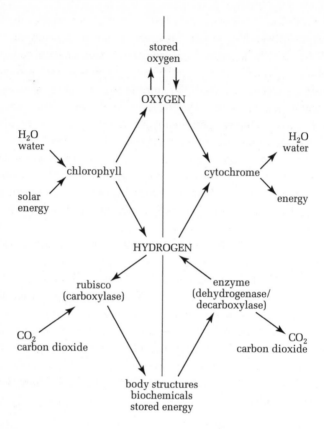

Figure 5.1 The balance of the twin activities of life: synthesis and oxidation. The page is divided down the centre. On the left, photosynthesis uses solar energy to split water into hydrogen and oxygen. The hydrogen is used to 'reduce' chemical compounds, making them energy rich and enabling them to be assembled into the structures of life. On the right, hydrogen is combined with oxygen to form water and release energy, as in a flame. The hydrogen is extracted from the energy-rich compounds made on the left and carbon dioxide is released.

oxygen behind. The process, known as photosynthesis, requires an input of energy that plants obtain from sunlight. Water has obvious advantages as a hydrogen source because, apart from being universally available, the released oxygen can diffuse away as a dissolved gas. Alternative sources of hydrogen such as hydrogen sulphide, H_2S, are possible, but then solid insoluble sulphur instead of dissolved oxygen is inconveniently left behind, although some photosynthetic bacteria have learned to live with it. Invoking H_2S as the hydrogen source implies an unnecessary layer of complication because life, being water based and indeed immersed in the stuff for most of its existence, was never short of water; why go looking for H_2S? Water is what plants use now as a hydrogen source and there is every reason to suspect that the idea has been around for a very long time. The result has been a release of oxygen via the oceans into the atmosphere; as far as we know, this is where all atmospheric oxygen came from.

Plants not only photosynthesise and release oxygen. They also respire, which is the opposite, the use of oxygen to oxidise or 'burn' fuels to release energy and reform water (Chapter 2). The two processes of photosynthesis and respiration complement each other, working in mirror image (Figure 5.1). Photosynthesis harnesses solar energy to insert hydrogen into energy-poor carbon dioxide to make energy-rich carbohydrates, which may be assembled into bigger structures such as starch and cellulose. Respiration gets some of the energy back by breaking down energy-rich carbohydrates, extracting their hydrogen and oxidising it back to water. The energy retrieved in this way powers the plant's life processes. Carbohydrate has acted as a temporary energy store.

The conversion of oxygen and hydrogen to water is catalysed by an enzyme called cytochrome oxidase, at the end of a chain of haem-containing cytochromes that oxidise and reduce each other. Since rubisco could be 3.8 billion years old (Chapter 4), a similar antiquity for cytochrome is equally plausible. Geologists can say from the composition of ancient rocks that the atmosphere was devoid of oxygen 3.8 billion years ago. But if primordial microscopic life was sufficiently sophisticated to possess rubisco and fix carbon dioxide, there is no reason why it should not have possessed a cytochrome to gather some of the oxygen released by photosynthesis before it escaped from the cell (rubisco can also fix oxygen, but this is not central to the discussion).

This argument is rapidly escalating because the processes of photosynthesis and respiration require highly complex supporting machinery, just as protein synthesis did. Respiration (biochemically this is the conversion of hydrogen and oxygen to water, not mechanical breathing) is a rich source of energy equivalent to burning hydrogen and oxygen in a flame. Enzymes that catalyse a reaction as difficult as respiration would never have developed and been retained without some selectional advantage. The very survival of life requires a continual input of energy because life's molecular structures are intrinsically unstable, tending to decay and lead to death. Energy for renewal and maintenance is required even by life that is virtually at a standstill or dormant.

For decades scientists were completely baffled as to how the cell captured the energy released when hydrogen and oxygen formed water. The trick was named oxidative phosphorylation because the cell was known to make ATP (AdenosineTriPhosphate) by adding another phosphate to ADP (AdenosineDiPhosphate), a thermodynamically unfavourable process that harnessed the energy from making water – but how was it harnessed? For years they searched for hypothetical carrier molecules that were supposed to gather up little bits of the energy and put them all together.

The problem was cracked in the 1960s by Peter Mitchell, aided by Jennifer Moyle, for which Mitchell received the Nobel Prize in Chemistry in 1978. Mitchell was an unusual character who left a career at Cambridge and Edinburgh owing to ill health and continued research at a small research institute that was effectively his own, known as Glynn Research, in Cornwall, Britain. Mitchell's stroke of genius (as is often the case) was to show that the process was philosophically simple, whereas numerous other researchers had been competing to make it more and more complicated. The story is fascinating and the lesson to science is that if hundreds of people have been searching for something – the mythical energy-storing proteins known as coupling factors – for 25 years without the slightest sign of success, and the list of specially contrived excuses for not finding them grows by the day, then the first bright spark to hit on the idea that they don't exist may well be on to something. The parallel with Einstein's line of thought a half-century earlier is unmistakable: if every attempt to measure the ether wind has failed, maybe there's no ether wind.

There's another interesting parallel between Peter Mitchell and other scientists who have broken fresh ground: he blended together knowledge gleaned from different fields in a unique way. His postgraduate research on the action of newly discovered penicillin gave him a detailed insight into the structure of cell membranes (where penicillin acts). But he also had a mastery of thermodynamics and a flair for constructing electronic apparatus. All of these played a part in unravelling what he termed the chemiosmotic hypothesis of oxidative phosphorylation (oxidative because oxygen converts hydrogen to water, phosphorylation because another phosphate is incorporated into ADP to make ATP).

In outline the principle is simple. Cells contain mitochondria, small bodies rather like bacteria (indeed they may be descended from bacteria; Chapter 7). In oxidative phosphorylation the energy released by the synthesis of water inside the mitochondrion is harnessed to eject hydrogen ions (protons) from the mitochondrion. Protons, akin to an electric current, flow back in, but pass through an enzyme of many parts known as ATP synthase, embedded in the mitochondrial membrane. The flow of protons powers the combination of ADP plus phosphate to make ATP, thereby storing energy (momentarily). The membrane is essential for maintaining this proton difference, or gradient between the outside and the inside, to drive the ATP synthase reaction. (The returning protons are not the same ones, of course; protons are everywhere, and tend to replace the deficiency in the mitochondrion.)

The ATP in turn drives the synthesis of other molecules such as proteins, and powers mechanical activities such as swimming. ATP is somewhat less magic than is sometimes claimed. Life need not have selected ATP as a major energy currency but did; however, much of its effect depends on mundane concentration ratios rather than any chemical wizardry.

Bacteria do not contain mitochondria but perform oxidative phosphorylation as though they are mitochondria. Protons are pumped out of the bacterium and then flow back through the ATP synthase sites in the membrane, driving ATP synthesis.

The ATP synthase is a stunningly intricate piece of machinery that actually spins in the manner of a molecule-scale electric generator – real nanotechnology. It can even spin backwards and reverse the reaction! The difficulty with projecting oxidative phosphorylation back 3.8 billion years is only subjective: how could such brilliant nanotechnology have existed so long before humans coined the word? How could oxidative phosphorylation function with any significant bits missing from this infinitesimal machine? Objectively, there is no reason why it should not have existed that early. The evidence for carbon dioxide fixation and photosynthesis existing in cellular life 3.8 billion years ago demolishes any argument that the biochemistry of the time was anything less than comprehensive. Since photosynthesis is at least as intricate as oxidative phosphorylation, life sufficiently advanced to have developed the one could have developed the other.

It's difficult to see how plant life, even as primitive as photosynthetic bacteria, could exist without both photosynthesis and respiration. Modern anaerobes that (by definition) do not respire do not contradict this assertion because they depend on a supply of ready-reduced food manufactured by other forms of life. Oxidation as a chemical reaction does not necessarily involve oxygen. Anaerobes extract energy, albeit inefficiently, by converting food to end-products still rich in energy, such as alcohol (which will still burn!). The fermentation of sugars by yeast to make wine is a good example, and anyone who has done it will recall that a bubbler is fitted to prevent oxygen from getting into the brew. The wasteful inefficiency of anaerobic metabolism, with energy-rich methane or alcohol being thrown away, would have provided a powerful incentive for primordial life to adopt oxygen for oxidation.

Clearly, life is not simple. Selective membranes, ribosomes that synthesise proteins, the metabolic pathways supplying the cell's never-ending demand for amino acids and other simple chemicals – these do not add up to anything like an explanation of life in molecular terms. The solution to how life conserves the energy from the synthesis of water was worth a Nobel Prize, but that process is in the present. It is happening all the time and is synonymous with life. What we really want to know is how such a complex process put itself together in the remote past.

WATER: BREAK IT OR MAKE IT?

Which came first, photosynthesis or respiration?

Irreducible complexity again! The problem is undercut by answering that both appeared at the same time. Primitive life developed around an ability to utilise solar energy. The distinction between aimless chemistry and life is that in life the energy is channelled into doing something subjectively useful. It went in the direction of synthesising energy-rich carbohydrates. From our perspective, that has two immediate positive benefits. First, carbohydrates are building blocks for making plant cell walls, DNA, RNA, hydrogen-carrier molecules and many other things.

Second, carbohydrates are an energy resource available for re-oxidation on demand. Hence sugar in the coffee!

The mirror-like interdependence of photosynthesis and respiration is a clue to their likely origins in a single molecule. As soon as chlorophyll and its associated photosystem had gained the ability to split water, there existed the twin opportunities for utilising both hydrogen and oxygen. Chlorophyll and cytochrome have a great deal in common. Both are proteins with an attached haem-like porphyrin ring in which metal atoms act as a temporary store of either electrons or oxygen. The similarity of chlorophyll and cytochrome may not be a coincidence if they had a common origin in a molecular partnership between a protein and a porphyrin ring. The differences are in the side chains attached around the ring, and in chlorophyll having magnesium where cytochrome has iron. A striking similarity between modern chlorophyll and cytochrome oxidase (the particular cytochrome that utilises oxygen and also contains copper) is the long tail on one corner of the porphyrin ring, containing 23 carbon atoms in chlorophyll or 17 in cytochrome oxidase; both also have methyl ($-CH_3$) side groups.

Success has been infectious in biochemical evolution and the porphyrin–protein theme has been capitalised on to build not one but a chain of consecutive cytochromes feeding their electrons into cytochrome oxidase, where with protons and oxygen water is formed (hydrogen is an electron and a proton). Significantly, these successive cytochromes have porphyrins a little different from those in chlorophyll and cytochrome oxidase in omitting the long tail – and so does haemoglobin, which leads on to the story of how oxygen can be kept.

OXYGEN: LOSE IT OR USE IT

The problem with the story so far is that life demands a supply of oxygen available instantly on demand, otherwise respiration stops. That's easy enough nowadays with an oxygen-rich atmosphere, but before the atmosphere contained much oxygen it would have been difficult. A better arrangement would have been to stockpile oxygen in the cell ready to use. This would smooth out supply and demand, enabling a measure of control so that ATP synthesis could be regulated according to the organism's need for energy. This idea is heading in the direction of haemoglobin, a logical adaptation of a cytochrome, complete with the familiar iron–porphyrin ring but with shorter-style side chains. (Strictly it should be called a myoglobin if concerned more with oxygen storage than transport, but myoglobins are haemoglobins and the general term haemoglobin is widely used.)

For years biochemists puzzled over the existence of haemoglobin in creatures too small to need it, and why it was needed over a billion years ago. Why should tiny organisms or single cells without a true circulation possess haemoglobin when they could obtain oxygen by diffusion? Why should an oxygen transporter date back to a time when the atmosphere and oceans provided little oxygen to transport? The answer may lie in the question: the near-absence of oxygen from the primordial atmosphere made it valuable to have a scavenging molecule, a haemoglobin precursor, that could trap meagre oxygen supplies released by photosynthesis.

The history of oxygen in the atmosphere is very much disputed. Many authorities place the major period of oxygen enrichment around 600 to 1000 million years ago, based on geological evidence, indirect deductions from geological sulphide deposits, and the slightly elliptical premise that the Cambrian explosion of evolution was facilitated by the increased oxygen supply. A fair consensus is that a billion years ago, the atmosphere had less than a hundredth, and two billion years ago about a thousandth, of its present oxygen. Others say that as long ago as 1.5 billion years the oxygen in the atmosphere had risen to a level that could be called aerobic. At times the percentage of oxygen may have been higher than the present 21 per cent, peaking at up to 35 per cent around 300 million years ago according to Robert Berner, which could have had a dramatic effect on evolution (Chapter 10).

Some geological evidence suggests that the first traces of oxygen appeared in the atmosphere more than 2 billion years ago. If it was all produced by photosynthesis as the evidence suggests, the productivity of photosynthetic life at that stage exceeded the combined capacity of life, oxidisable minerals and the oceans (in which life then lived) to absorb it. Tight reasoning by Jochen Brocks and his team about organic molecules preserved in ancient rocks (mentioned in Chapter 4) supports the concept that photosynthesis was not only generating oxygen 2.7 billion years ago but was established and sophisticated. This is evidence of a different kind from ^{13}C depletion that indicates that life existed well over 3 billion years ago (Chapter 4).

An ancestral oxygen-scavenging biomolecule of high affinity could have trapped much of the oxygen in the biosphere rather than it escaping to the oceans and atmosphere – up to a point. Regardless of the number of such scavenging molecules in the cell and how closely positioned they were to the site of water splitting, inevitably some oxygen would escape. For a long time the equilibrium may have kept the atmosphere virtually anaerobic. (Anaerobic, strictly meaning a lack of air, is not as good a word as anoxic, meaning zero oxygen, or hypoxic, meaning low oxygen, but is widely used.) This is another example of how the second law of thermodynamics favours life rather than disfavouring it (Chapter 3; explored in more detail in Chapter 6). The law says that disorder must continually increase in the universe as a whole (whereas life represents order). But the corollary is that molecules are in constant motion, and oxygen, for example, will sooner or later react with an oxygen scavenger and be captured. The massive size of early organic rock deposits indicates that the biosphere accounted for a considerable reserve of oxygen.

The role of haemoglobin appears to have adapted itself from a primordial storage capacity, when oxygen was scarce, to a reversible transporter function when oxygen became more abundant. The ability of haemoglobin to transport oxygen around a body contributed ultimately to human evolution by vastly increasing the possible size and metabolic activity of animals, which in turn supported the development of large solid organs such as the brain, leading to consciousness and intelligence. Modern vertebrate haemoglobin still retains its primordial storage function in the sense that the blood contains a store of oxygen that is only released to the tissues on demand. Haemoglobin does not push oxygen into the tissues but holds it until called for by cytochrome oxidase, which in turn responds to the demands of the cell for energy. The whole process is driven by energy demand, not supply.

Reversibility (and another layer of sophistication known as co-operativity) is what makes haemoglobin an efficient oxygen transporter. A non-reversible oxygen carrier would uselessly retain its oxygen and supply nothing to the cells. Reversibility distinguishes haemoglobin from simple iron, which is easily oxidised, but irreversibly so, the way rust forms. Haemoglobin contains iron to carry oxygen, true enough, but the convoluted structure of the complete molecule functions not to enable but precisely to prevent an irreversible union between the iron and the oxygen.

Many invertebrates have a haemoglobin gene and a number of them express the gene to make haemoglobin (species do not necessarily express all their genes, at least not all the time). Haemoglobin is widespread throughout the animal kingdom and, curiously, in some plants. The haemoglobins of non-vertebrates are similar to those of vertebrates but are stacked together differently. The unavoidable conclusion is that the gene for such a complex protein was present in a very ancient organism to which most of life can trace its ancestry.

There cannot be complex proteins without the entire panoply of other complex proteins (enzymes) to catalyse their assembly and the genes to code for them. Specialised proteins existed as far back as sequence comparisons allow us to see, and less direct but extremely powerful evidence points towards many of the enzymes for cellular photosynthetic life being in existence 3.8 billion years ago. It stands to reason that cells would not have directed their ingenuity to the manufacture of a select few prominent proteins in ridiculous isolation, but must have made the minimum set required for life, generally thought to be around 1000 to 1500, depending on how genuinely complete and independent the life was.

LOSE–LOSE PARADOX

The conclusion is inescapable. Life at least 3.8 billion years ago was biochemically almost as complex as it is today. That date is a mere 2.5 per cent away from the 3.9-billion-year age of the Earth's solid surface; the dates are scarcely distinguishable statistically. (The earlier date of 4.1 billion years for a surface suggested by recent lutetium–hafnium measurements would drag the antiquity of life with it. It would not increase the window available for life to originate or change the argument.)

The figures must be interpreted with latitude. Suppose that there might have been 100 million years between the solidification of the Earth's surface and the earliest apparent life. That is still a very long time. For comparison, the demise of the dinosaurs was only 65 million years ago. But as the date of the earliest life on Earth converges on the date for the formation of a surface, some rather awkward possibilities loom.

Any notion that life originated on Earth is trapped in a lose–lose paradox. If, in the beginning, life took very little time to develop from nothing, then a great deal must have happened rapidly. That implies that the process was relatively easy and capable of happening at countless locations throughout the universe, all the time, automatically undermining the idea of its being special to Earth. It would be very odd if life developed on Earth in a period representing little more than 1 per cent of the universe's existence but nothing had happened anywhere else in the preceding 99 per cent. However, if life inside the cell has developed little over detectable time, as appears to be the case, then change has been exceedingly slow – which does not sit comfortably with the idea of a rapid origin. It would be as though at least 95 per cent of the work of life's origin and evolution happened in the first 1 or 2 per cent of the Earth's existence, which doesn't make a lot of sense – unless life on Earth received a considerable head start. If life coming into existence is an inherently unlikely and exceedingly slow process, then the necessary timescale was not available on Earth. Either way, we are directed into an era long before the formation of the Earth.

THANKS TO THERMODYNAMICS

If life was never simple, how did it start?

Up to this point in the book there have been many mentions of *cellular* life and of life *on this Earth*. Primordial life on this Earth was cellular and complex. Life might possibly have originated on Earth, which must mean sometime in the past 4 billion years, but there is no reason why it should have. Why impose artificial and unfounded constraints on where and when? Even the simplest conceivable cellular and photosynthetic life may have taken billions of years to reach such a degree of complexity, extending back to long before the Earth was thought of. Much more likely, life came together not on Earth, and not in the past 4 billion years, but throughout all the space and time of the universe.

The conundrum that life can never have been simple but could not have started off complex – so-called irreducible complexity – is another red herring There was no moment after which life existed and before which it didn't. There was no place, no time, no demarcation between life and the chemistry that led up to it. Life has always been forming, everywhere. What we see now, we recognise as life. What a time machine would reveal about the unimaginable past might not seem to be life. This chapter often uses the Earth as a convenient frame of reference, somewhere familiar for purposes of comparison or calculation, but the discussion is not only about the Earth. It is about how the reactions of life could have accumulated on numerous planets, in their atmospheres and indeed in space itself.

Life was pushed from the bottom rather than pulled from the top. Unless one wishes to move outside the bounds of science, it is important never to think of life as needing to function in any particular way or heading towards any inevitable target. There is no target and never has been – least of all the human species. Life is its own record of how it developed and of those random experiments in its history that were successful. There is, however, a subtle distinction between random and haphazard. Life is not a haphazard collection of biochemical reactions, a haphazard mess of limbs and organs. Far from it. Every living cell, each living species (except a few varieties bred by humans!) is a beautifully tuned and integrated machine structurally and functionally – it must be so to survive in its environmental niche – but the mutations that shaped its history struck randomly.

Whether life emerged as a compendium of chemical adventures recruited from the universal primordial soup or within some isolated planetary puddle, somehow the bits had to come together. The laws of thermodynamics (simply meaning heat and movement) provide the crucial explanation of how mixing occurs over large distances.

THREE SIMPLE RULES

It's rather odd that people who set out to deny the origin and evolution of life often try to use the principles of thermodynamics to bolster their argument. The favourite ploy is to make the blindingly obvious statement that all things eventually fall to pieces and living things die, with the implication that the reverse cannot happen. In other words, dead chemicals could never have come alive. This strange reasoning is entirely perverse because it ignores the equally self-evident fact that things, including life, exist before meeting their demise. They exist because of, and not in defiance of, the laws of thermodynamics, which are in fact the key to the origin, evolution and all activities of life. There are no exemptions.

The falling-apart argument is supposedly based on the second law of thermodynamics, but it's best to take a simple overview of all three laws to see how they fit together (they have been mentioned already in Chapters 3 and 5).

The laws of thermodynamics really are beautifully simple. They are often distilled into three truisms: (1) you can't win; (2) you can't break even; and (3) you can't stop playing the game. Or in another idiom: (1) you can't heat the house without burning fuel; (2) no boiler is 100 per cent efficient; and (3) warmth will never completely go away − neither will the laws of thermodynamics.

The laws do not work against the origin and existence of life but overwhelmingly in its favour, so it's worth investing just a little time in getting a feel for them. The trick to a simple understanding is to realise that although thermodynamics is to do with energy moving around, the visible effect is usually atoms and molecules moving around. The link between the two is that released energy finishes up as heat, and the effect of heat is to increase the random motion of molecules (kinetic energy). Molecules consist of atoms joined together by bonds and the energy source we are talking about is the energy needed to form those bonds in the first place, and conversely released when the bonds are broken. Statements of the laws of thermodynamics from different sources can sometimes appear to be quite different, depending on whether they are framed to say something about molecules or about energy, but it's actually easier to think of both at the same time. It also makes more sense to visualise the happenings in a large population of molecules, not just one molecule.

A little more formally, the first law says that energy cannot be created or destroyed. When a spontaneous chemical reaction occurs, such as burning or oxidation, the starting molecules contain energy in their chemical bonds. So do the product molecules, but less, because some energy is lost in the reaction. That energy has to go somewhere and finishes up as heat (occasionally as a flash or a bang, but that ultimately finishes up as heat). The heat makes the product molecules hotter, so their random motion increases, and even as they cool they emit radiant heat that warms other things around, so the energy is never lost to the universe. The overall lesson is that the total energy in molecular bonds to start with is exactly the same as that present in different molecular bonds, and heat, at the end.

Sometimes reactions go the other way and an input of energy (such as from hot water) can lead to products containing more energy than the starting chemicals. But of course the water had to get its heat from somewhere, such as burning gas, which gets us back to the first example. A reaction that generates heat is needed to drive one that requires heat.

The practical effect of the first law on life is that living organisms need an input of energy when being put together and a continual input of energy to stay alive. Since you can't get something for nothing, energy must be obtained by converting it from another form, namely food. The food at the start of the food chain was made by plants, requiring a lot of energy, which was radiated by the Sun.

The second law is the one hilariously misrepresented by people trying to deny evolution, yet is exceptionally easy to grasp and is so intuitive. You boil a kettle of water and leave it. A couple of hours later it's cold. In obedience of the first law, the heat that disappeared from the water has not disappeared from the universe, it has merely made the kitchen a tiny fraction of a degree warmer. Now for the second law. This simply says that you cannot push the warmth from the kitchen back into the kettle. You might think you could: for instance start up an air conditioner, cool the kitchen and use the heat extracted to warm some water.

Indeed you could, but it would cost more energy than you would recover. Not only can you never win (first law), you cannot break even (second law). (Heat pumps do work, quite well in fact; but they consume electricity and the energy output is not free.)

The second law is usually expressed as: *in a defined system*, the total amount of disorder (given the name *entropy*) always increases. A defined system is any system or place you care to draw a line around: a room, the Earth, the solar system or the entire universe. Entropy, in effect, is energy that still exists but is no longer available in practical terms. A defined system is for you to choose, provided that you define it clearly. Order in the above example is having all the heat in one place, namely the kettle of hot water; disorder, or entropy, is the same heat scattered throughout the kitchen, the house and the garden. Self-evidently, what starts off as order heads towards disorder.

The special significance of the second law to the origin and evolution of life is that it also affects the motion of molecules, not merely energy. Drop a tea bag into a cup of water and watch (assume no stirring or convection). Brown molecules slowly diffuse from the bag, spreading through the water. At first there is a clear gradient of darker colour closest to the bag, fading with distance. The higher the temperature, the more kinetic energy the molecules have and the faster they move (so to make iced tea it's better to make it hot first, then ice it). Eventually the colour will be evenly distributed throughout the cup; there will be equilibrium. What started off as order, with all the colour in the bag and none in the water, eventually finished up as disorder, with the colour everywhere. As with the heat from the kettle, the colour cannot be pushed back into the bag (well, you could boil away the water and retrieve the miserable tea bag and what came out of it, but at great energy cost).

The consequences of the second law for both the origin and the existence of life cannot be overstated. Just as the tea spreads throughout the cup until it is everywhere, all the chemicals in a primordial soup will continually spread out and mix well. The second law of thermodynamics stirs the system, whether it be an ocean, a solar system or a galaxy.

Or a cell. Even within a cell the molecules necessary for life are in rapid motion, colliding and interacting. This is good and bad: good because a molecule needed by a particular mechanism of the cell will find its way there rapidly, intracellular distances being very short; but bad because the tendency to disorder means that the cell spends a lot of energy fighting to lock up special molecules where they belong, such as inside mitochondria. The same tendency to disorder ensures that the DNA-duplication mechanism is continually bombarded with the wrong base-containing nucleotides and occasionally lets the wrong ones in, causing the mutations without which evolution cannot happen (Chapter 3).

The endemic hang-up that some people have over the second law is a fixation on disorder always increasing, while ignoring the essential proviso that it applies only to an isolated system. The tiresome argument is that living organisms are highly ordered (true) and the universe is becoming increasingly disordered in accordance with the second law (also true), therefore disorder could never have been reversed to achieve the order essential for life (nonsense). As with getting the tea back into the bag, or distilling brandy from wine, all that's needed is a supply of energy. Certainly the universe as a whole is becoming more disordered, but there is not the slightest objection to energy from one system (the Sun) driving an increase in order in another system (life).

Another consequence of the second law is that perpetual motion machines are completely impossible – just forget the idea. Energy becomes degraded and less useful with every conversion, so you can't get back as much as you put in (the second law), let alone more (the first law). Life that needed no energy or food would be a perpetual motion machine and is thus impossible. Life cannot stop using energy, because the slide towards disorder and decay is immediate. For the same reason, a dead cell cannot be brought back to life. Hibernation is possible because life then continues at a low level, but completely suspended animation is impossible. While frozen cells or dried spores might last an extremely long time, they must eventually die.

The third law of thermodynamics acknowledges that a temperature as low as absolute zero is unattainable. Absolute zero, the theoretical bottom of the Kelvin scale, would be −273.16 degrees on the Celsius scale. This is a temperature that has been approached experimentally to within a billionth of a degree, but cannot be reached. The third law has something positive to say about the origin of life because the temperature of any matter in space must be above zero Kelvin, if only 2 or 3 degrees. That being the case, all matter has thermal motion, will spread around the universe and is not exempt from chemical reaction, albeit exceedingly slow. Conversely, chemicals initially created in abundance in hotter regions of the universe can be preserved for great periods of time and transported vast distances at very low temperatures.

CHEMISTRY: WHAT ELSE?

The seventeenth-century philosopher Johann Baptista van Helmont planted a willow seedling in a large pot, watered it and watched it grow for five years (this was before the days of research grants). The great classic experiments in science have usually been so beautifully simple that the only outstanding question is why they were not done before. Van Helmont's experiment was no exception. He was one of those wonderful scientists who had an interest in all sorts of things, including biology, chemistry, air and gases, and also accurate weighing, all of which were brought together in this famous experiment. After five years he unearthed the tree and found that it weighed (in modern equivalent) about 164 pounds or 74 kilograms. The soil had lost very little weight and van Helmont's disarmingly straightforward deduction was that the tree had created 164 pounds of matter out of nothing but water.

Not quite! We now know that atmospheric carbon dioxide would have contributed the carbon and much of the combined oxygen, and that the soil must have contributed nitrogen and some trace minerals, although the slight apparent decrease in soil weight was probably attributable more to a moisture difference and incidental losses. Nevertheless, the general principle that an advanced form of life required nothing material except water (and air) to construct itself was remarkably close to the truth. There is no reason to suppose that primordial life was any more demanding than van Helmont's tree. Quite the opposite: water and air are good enough now and were good enough then.

The foundation reaction of life may well have been the splitting of water with solar energy and a catalyst (Chapter 5). At a stroke that provides life's two absolute essentials – reducing power for growth (hydrogen) and oxidising power for activity (oxygen). As we have seen, the short list of indispensable chemical elements necessary to transform that rudimentary reaction into a working model of life is very short indeed: mainly carbon, hydrogen, nitrogen, oxygen and phosphorus, plus a little magnesium, iron and copper. Common elements such as sodium, potassium and calcium are invariably present, but whether they would have been essential is a different question. The universe is mainly hydrogen, but these other elements, including the metals, are relatively abundant. Logically the materials and reactions captured by life were those readily available, and more esoteric ways of achieving the same objectives, while undoubtedly possible, stood less chance of adoption.

Life had to start somehow and it could only have been based on an accumulation of chemicals formed spontaneously. Which pre-life chemicals would form spontaneously? Many planetary atmospheres (including the early Earth's) would have contained water vapour (H_2O) and at least some hydrogen, nitrogen, nitrogen oxides, CO_2 (carbon dioxide), NH_3 (ammonia), CH_4 (methane) and HCN (hydrogen cyanide). Vapours such as formaldehyde (HCHO) and methanol (CH_3OH) may have been present. Sulphur, phosphorus and chlorine are likely to have been present as H_2S (hydrogen sulphide), SO_2 (sulphur dioxide), CS_2 (carbon disulphide), PH_3 (phosphine), HCl (hydrochloric acid) and $COCl_2$ (phosgene).

Under atmospheric conditions these chemicals can react and combine to form small biological components. In 1913 Walther Löb demonstrated the synthesis of the amino acid glycine in a mixture of ammonia, carbon monoxide and water subjected to an electrical discharge, but he was interested from a chemical point of view and probably had no great interest in the origin of life. In 1953 graduate student Stanley L. Miller at the University of Chicago looked at the question with thoughts about the possible origin of life. Miller was inspired by the ideas of his supervisor Harold Clayton Urey, who in 1934 had won the Nobel Prize after discovering heavy hydrogen (deuterium) and heavy water. They wanted to know which chemicals would appear in a primordial Earth atmosphere, so Miller did the logical thing. He tried it.

Miller's apparatus subjected a best-guess primordial atmosphere, including water vapour, to high voltage sparks simulating the effects of lightning. The experimental atmosphere contained nitrogen, ammonia, carbon dioxide, carbon monoxide, methane and hydrogen (no oxygen). After heating, recirculating and flashing 60 000-volt sparks through this mixture for several days he analysed the liquid and found a surprisingly comprehensive crop of small organic molecules. Notable among them were four of the amino acids common in proteins: glycine, alanine, glutamate and aspartate, and some other amino acids not usually found in proteins. Although the apparatus is relatively simple, much of the achievement lies in the analysis of the products afterwards.

Amino acids were not the only molecules generated in Miller's primordial soup. Other biologicals found included formic, acetic, propionic, lactic and succinic acids.

Needless to say, the experiment has been repeated many times since, and work by NASA along the same broad principles has simulated reactions occurring in sparse gas clouds in space. Even with the relatively mild conditions used by Miller the range of products is impressive, and with more modern detection additional amino acids including proline and valine are usually found. Those six amino acids, creatable out of nothing but an atmosphere and energy in a matter of days, contribute several key properties to protein structure: valine is hydrophobic and helps to form the core, glutamate and aspartate are hydrophilic and tend to be on the surface, while glycine and alanine are small enough to fit in most parts of a protein. Proline has special properties that induce bends in proteins, as does glycine. By adapting the experiment it has been possible to generate the nucleic acid bases A and G, and more recently C and U, the components of RNA (Chapter 5).

The Earth has other sources of energy than lightning in its repertoire. Molten volcanic lava is not only very hot but contains a wide variety of mineral compounds that might act as catalysts. Edward Anders has studied the effects of simulated hot laval constituents on a primordial atmosphere and achieved a list of products surpassing Miller's, including additional amino acids and nucleic acid bases. A molten surface and a hot atmosphere would have created a rich catalogue of chemical compounds and polymers. Before the formation of the ozone layer (which required oxygen) the Earth would have been bombarded with ultraviolet solar radiation, one of the effects of which is to encourage polymerisation, the formation of the long chain molecules characteristic of life. Natural radioactivity in mineral surfaces, particularly from uranium and thorium, has also been suggested as a polymerising influence.

If life were to develop in the swirling clouds of a dense planetary atmosphere, it would benefit from the catalytic effects of metal ions such as are often involved in enzyme reactions. Less obviously, heavy elements and metals would be available in a planetary atmosphere because gaseous compounds contain them. Elements such as selenium can exist as the gaseous hexafluoride (F_6Se) and dihydride (H_2Se); there are many parallel examples. Heavy metals important to life including iron and copper can form volatile carbonyls with carbon monoxide under the catalytic influence of molten rock. Sodium and similar elements can exist as atmospheric vapour. Another significant source of airborne metals and minerals, including notably magnesium and iron, is the vaporisation of asteroids and cometary material. The amount of material hitting the Earth's atmosphere is this way is about 40 000 tonnes per year.

Modern experiments such as Miller's to make biochemicals from atmospheres are confined to an infinitesimally short timescale in evolutionary terms: days instead of millions of years. Doubtless many other compounds are formed in traces that would become detectable, given much longer. The crucial point is that on a primordial planet these chemicals would rain down into lakes and oceans to accumulate in ever-increasing concentrations. They would not be consumed except by decomposition or by reacting together (thereby making more chemicals). Today, on Earth, a lake of organic chemicals would very quickly be colonised by bacteria and other forms of life. But we are not talking about Earth any more, or any one planet. Primordial soup has a more general meaning, of all soups on all planets throughout the history of the universe, and loosely includes atmospheric clouds and interstellar material. On a lifeless planet over millions or billions of years one could expect vast quantities of organic chemicals to accumulate, for the simple reason that there is nothing to eat them. Primordial soup had a decent shelf-life. A planet without life is, by definition, sterile, and novel combinations of simpler chemicals would continually increase the variety of molecules present.

At least, that's almost true. If certain individual reactions that we now associate with life were present in a primordial soup, discrete processes such as fermentation and digestion could occur. But just as pre-cellular life would itself be too slow to recognise, so would pre-cellular fermentation reactions. Therein, however, lies the answer to the paradox of irreducible complexity – if life must be complex, how could it ever have been simple? There is no problem and no paradox of irreducible complexity except that the definition of life has simply been set too high. The working definition of life implies cellular life, controlled by the equivalent of a nucleus, capable of reproduction and other activities. Of course, this did not suddenly jump into existence. Prior to that there would merely have been chemical reactions, or sets of reactions such as those contributing to fermentation, happening vanishingly rarely and unconfined by cells, because the right molecules would collide infrequently.

No doubt the initial stages in the formation of life happened infinitesimally slowly. Some might equate that with zero and take it as proof that they did not occur by chance at all. The other way of looking at it is to say that the more unlikely the reactions, the slower the process of synthesis, the longer it must have taken and therefore the earlier it all began – much earlier than the Earth.

LIVING ON THIN AIR

The idea of life being sustained by nothing much more than atmospheric components might sound unconvincing, yet as van Helmont showed, plants do it all the time. Plant life clearly demonstrates the extraction of its materials from the atmosphere by growing essentially on carbon dioxide and water, plus a nitrogen source, a few trace minerals and sunlight. The only real problem is the nitrogen source, for which life has relied mainly on the legacy of nitrogenous chemicals recycled from previous life (until the invention of manufactured fertilisers). Ironically, the Earth's atmosphere now contains far more nitrogen (about 79 per cent) than anything else, but for this to react biochemically is a major problem. Nitrogen atoms are held together in pairs, as N_2, by a powerful triple bond that requires considerable energy to break it. Only a few bacterial families can achieve this very first step in the reaction, splitting N_2, after which it can be incorporated much more easily into more reactive chemicals such as ammonia, and from there into amino acids, nucleic acids and other biological molecules.

The bacteria capable of 'fixing' nitrogen in this way are cyanobacteria such as *Oscillatoria, Rivularia* and *Nodularia*. Some lichen species superficially appear to fix nitrogen, but the true ability resides in cyanobacteria living inside them symbiotically. Leguminous plants, such as clover, have captive bacteria that convert atmospheric nitrogen into ammonia for them. Growing legumes and then ploughing them into the ground is a convenient way of putting nitrogen-rich fertiliser into the soil.

Since nitrogen fixation is achieved by these simple root bacteria and some modern cyanobacteria, and organisms resembling cyanobacteria seem to have appeared extremely early in evolution (see the latter part of Chapter 4), one might surmise that nitrogen fixation was one of the earliest mechanisms of life. There is no reason why this should be, however, because the problem was not there in the first place. The primordial atmosphere would have contained non-biological ammonia (NH_3), which life can easily utilise by converting it to the amino acids glutamate and glutamine. One source of atmospheric ammonia would have been the chemical combination of nitrogen and hydrogen, catalysed by minerals at the high temperatures and pressures associated with deep-sea hydrothermal vents, which were probably extensive on the early Earth. Lightning and other sources of energy are likely to have stimulated the synthesis of amino acids, amino sugars and the nitrogenous bases of DNA and RNA.

Only after ammonia and oxides of nitrogen had dwindled from the atmosphere would it have become beneficial for alternative sources to come on-stream. Prior to that the difficult biological conversion of nitrogen to ammonia was unnecessary, and indeed it can be performed today by only a select few bacterial species, which makes them not so simple after all.

Down to Earth (or Somewhere)

Atmospheres and oceans may be good places for many of the reactions of life to develop, but mineral surfaces add a new and crucial dimension. They provide catalytic surfaces and minute fissures, both of which allow localised concentrations of chemicals to build up. This is advantageous, because higher concentrations of chemicals react faster. An advantage of biochemicals being washed into a lake of primordial soup is that evaporation would have the effect of raising concentrations, dramatically so if a large lake evaporated to a small volume. This is precisely what seems to have happened on Mars, where the most recent exploration suggests that original oceans have evaporated to the point where no free surface water remains.

The importance of chemicals being concentrated together can be seen by analogy with balls on a pool table. If you fired a ball at random across a table with only one other ball on it, the probability of hitting another ball would be almost nil. But if you fired your random shot and the table had 1000 balls on it, the probability of missing another ball would be virtually nil – in fact there would be numerous rebounds. Chemical reactions occur not because two of the right molecules are somewhere in the soup, but because a lot of them are in close proximity – the concentrations are high.

The single most pivotal event in the transition from a primordial soup to life could well have been a partitioning into small compartments by insoluble organic polymers. Previous to that stage, the soup contained such a comprehensive assortment of organic chemicals and catalytic particles, accumulated over aeons, that the individual reactions of life were already happening. Diffusion and mixing would have ensured that the ingredients of the soup were uniformly distributed such that a few molecules of any particular chemical were never far away. This is the second law of thermodynamics in operation. Without it, a repertoire of different chemicals produced in numerous crevices of a primordial planet would have stayed where they were formed, without meeting. Fortunately, the second law dictates that local concentrations would spread out endlessly, which is exactly what was needed. You can pour a bottle of milk into the sea, but you can't pour it back out. Ultimately, the primordial soup was a mixture of everything that poured into it, providing everywhere the rich variety of molecules necessary for life to proceed.

Life's brilliant but unwitting trick was to partition the primordial soup into microscopic volumes. The trick was not to try to embrace the entire planetary stockpile of ingredients, but to do the opposite and confine a representative mixture to a very small volume, where concentrations and reaction rates slowly built and reinforced each other. As partitions grew progressively smaller, the critical ability to sustain high concentrations of metabolites was achieved more easily until the natural size range of cells was reached, which logically is the volume readily controlled by one set of genes.

Much is made of the role of water as the essential basis of life, which is correct. But life devotes considerable energy to the task of excluding excess water from cells, by a combination of strong cell walls and active pumping mechanisms. Another water-excluding trick is the polymerisation of sugars that otherwise would provoke an influx of water owing to osmosis (the tendency for water to flow into cells to dilute the contents). For example, glucose causes much less osmotic inflow of water if it is converted into starch because large numbers of small molecules (glucose) have been converted into a small number of large molecules (starch). Many species of bacteria, protozoa, plant seeds and even some relatively advanced multicellular animals can survive in the form of dormant resting stages by dehydration, but there is no organism that can do the converse and survive a massive dilution of its cellular contents.

Up to Speed: Cells and Enzymes

It's becoming clear why two major developments explained in Chapter 5, cells and enzymes, have endowed the reactions of life with the diagnostic property of speed. Of course, speed is a purely subjective criterion of life imposed by humans, always short of time and preoccupied with measurement and classification. In reality there was no qualifying rate of activity that defined the beginning of life, which is part of the problem of identifying some notional starting point for life. Both cells and enzymes have the effect of confining chemicals in high concentrations that greatly increase their probability of interacting. The chief advantage of cells is that all chemical reactions are driven by the concentrations of chemicals, not merely their ratios. A few of the right molecules scattered throughout a lake may never get the chance to interact. Confining these same molecules to a small volume will enable the reaction to happen.

A step in the direction of cell formation would have been the appearance of insoluble chemicals in the soup. As a rule, large molecules are less soluble than small ones, and some simple chemical features on molecules, such as $-CH_2-$ or $-CH_3$ groups, make them less soluble in water than do other groups such as –OH. Fatty acids (acids incorporated into fats) are a good example. Short fatty acids such as acetic and propionic acids (present in vinegar and certain less-memorable wines) are easily soluble, but longer versions such as palmitic and oleic acids (used to make soap) are fairly insoluble. As molecules in the primordial soup grew longer and more complex, the appearance of these insoluble, oil-like molecules would have been a certainty, whereupon they would have floated on the surface and formed an insoluble layer. Molecules of the fatty acid type in conjunction with glycerol and other small molecules tend to associate naturally in sheets about 7 to 10 nanometres thick and are the fundamental building units of cell walls and internal membranes.

A lake can be envisaged progressively filling with primitive organic chemicals as a result of the kinds of reactions highlighted by Stanley Miller and Edward Anders. A network of floating insoluble products builds downwards from the surface, trapping small pockets of water containing chemicals. Evaporation of the lake could play an important part in increasing the concentration of chemicals, leading to a sponge-like matrix of oily matter subdividing the solution into small pockets. The pores of the matrix confine small volumes of chemicals at high concentrations.

The reactions of life are virtually all catalysed by enzymes, which speed up each reaction maybe a thousand or a million times, in some cases more. Enzymes are proteins, some with non-protein adjuncts (but not all proteins are enzymes). Enzymes don't do the impossible, they don't make reactions happen that cannot happen anyway. And yet they do! Correctly stated, an enzyme only makes it possible for the reaction to run very much faster, so in theory at least, every one of life's reactions was capable of happening before its enzyme appeared. But by our impatient human standards many enzyme-catalysed reactions would happen so much slower without the enzyme that they would be virtually impossible. To say that an enzyme causes a reaction to happen is a biological colloquialism no more criminal than an astronomer commenting that the Sun has gone in.

Any explanation of the origin of life has to provide some reasonable account of where the thousands of enzymes came from. The processes catalysed by enzymes are all very simple reactions that happen under the mild conditions associated with life, supporting the idea that the reactions existed first and catalysis came along later. If an enzyme catalyses an existing biological reaction, there is benefit in retaining and refining it, but it makes no sense to adopt an enzyme for an unused reaction. The reactions of life were possible, at an infinitesimally slow rate, before they were speeded up by catalysts. Typical enzyme-catalysed reactions are the addition or removal of the components of water (H^+ and OH^- ions), the addition or removal of a phosphate group, or comparably simple processes. An enzyme that cuts proteins into pieces sounds rather exotic but the basic chemistry is simply hydrolysis, rounding off the broken ends with H^+ and OH^- as they are helped apart. It is the exquisite specificity of some enzymes, such as those capable of recognising and splicing special sequences in DNA and RNA (but still by simple chemistry), that tempts people to question how they could possibly have developed within any conceivable timeframe.

Enzymes, like all proteins, are huge molecules of extremely precise shape, and there are thousands of different types. Each protein is a long chain of amino acids joined end to end. Stanley Miller's experiment with the model primordial atmosphere generated several of the 20 different types of amino acid found in proteins and it is not difficult to conceive that, given millions of years rather than a few days, all 20 would have appeared. But this comes nowhere remotely near to solving the problem.

Now that life is here, it is very easy to observe how proteins are assembled. Each protein destined for a specific task is a unique sequence of amino acids linked in exactly the correct order. The protein then folds itself into a convoluted three-dimensional shape that has to be meticulously perfect. The slightest error, such as a wrong amino acid in one place, may distort the three-dimensional structure of the protein to such an extent that it fails to perform its function, and if that function is vital, such as life's functions tend to be (by definition!), then death results. Importantly, an incorrect amino acid (owing to a mutation in the gene) is sometimes tolerated since that is the key to evolutionary change (Chapter 3).

Working backwards from the completed protein, its unique sequence of amino acids is assembled inside the cell on an apparatus called the ribosome, using information carried by messenger RNA, which in turn carries the information imprinted in the gene DNA (Chapter 4). How could all this possibly have happened before primitive ribosomes, RNA and DNA sequences were invented? How could it have happened if any of these essential components were missing?

Ignore the Impossible

Authors dedicated to the cause of disproving the natural basis of life are extremely fond of a perfectly reasonable calculation, showing that indeed none of the known proteins and enzymes essential to life could possibly be recreated by the random joining of amino acids. They are quite right. Even if long proteins did form at random, there would be 20 alternative amino acids at every step in the chain and every protein molecule in the world would be completely different.

The calculation is very simple. There are 20 choices for the first amino acid in the chain (the primordial number may have been different, but it illustrates the calculation). Attach the second amino acid: the number of amino acid pairs is the permutation of 20×20, or 400. Add the third amino acid: now 8000 triplet combinations are possible ($20 \times 20 \times 20$). Add another: that's

160 000 possibilities for a chain of just four amino acids
(20 × 20 × 20 × 20). Every amino acid added to the growing chain
could be any one of 20 different choices. Now jump to a modest-
sized protein molecule containing 100 amino acids. The number
of different proteins possible is found by multiplying
20 × 20 × 20 ... a hundred times. Does that sound a lot? Oddly
enough, to a non-mathematician, 20 multiplied by 20 a hundred
times in succession is a large number but doesn't sound
astronomically vast; there would be far more grains of sand on
the beach, wouldn't there?

No, actually, there wouldn't! Just as 10×10 can be written as
10^2, a way of writing the number we are wrestling with is 20^{100}.
That's still slightly awkward because numbers consisting of 20
raised to a power are less familiar than 10 raised to a power, but
there is a simple way of switching between the two, so that 20^{100} is
in fact the same quantity as 10^{130} (in round figures; derived from
the fact that 20 is approximately $10^{1.3}$, and $10^{(1.3 \times 100)}$ is 10^{130}).
That's a figure 1 followed by 130 zeros.

Even 10^{130} is so easily written that it might not immediately look
astronomical, and yet it most certainly is; wildly, outrageously
gigantic beyond all possible contemplation. There are nowhere
near that many atoms in the entire visible universe, for which a
figure of 10^{79} or 10^{80} is often quoted (from considerations of the
volume of the universe and its average density of no more than 3
atoms per cubic metre; and if 10^{80} looks surprisingly small, that
illustrates the sheer incomprehensibility of such numbers). To
press the point, 10^{130} is larger than the number of atoms in the
universe – let alone grains of sand on the beach – by a multiple of
10^{50} *times*.

In case you've forgotten, we were talking about the number of
different possible proteins made of 100 amino acids, the answer
being 20^{100}, which is 10^{130}. A way of putting such a super-
astronomical number into perspective is to ask how long it would
take to make all the 10^{130} different possible combinations for a

chain of 100 amino acids, with 20 choices at each step. Assume that a protein is completed every second, each one different. The number of seconds in a year is about 30 000 000, or 3×10^7, so (ignoring the 3) the time required would be 10^{130-7} years, which comes to 10^{123} years. If that looks a bit depressing, go further and assume outrageously favourable conditions for random protein formation: the entire oceans of primordial Earth are stiff with amino acids, and different 100-amino-acid proteins are forming at the rate of one every second in every microlitre of ocean (about 1/30th of a drop). The volume of the oceans presently is about 10^{21} litres or 10^{27} microlitres, so all combinations of protein would be tested in 10^{123-27} years, which is 10^{96} years (ignoring duplicate proteins). If it happened every microsecond (10^{-6} of a second) it would still take 10^{96-6} or 10^{90} years. That is still 10^{80} *times* as long as the Earth has existed (which is less than 10 billion years, or 10^{10}).

That sort of calculation (with a variety of different numbers, they scarcely matter) is the thrust of a purported proof that life did not originate of its own accord and therefore originated of somebody else's accord. In other words, life was created supernaturally. The calculation says quite correctly that, given a pre-specified protein, there is not the faintest likelihood of its ever having fallen together by pure chance, not even once in all of time (to be fair, the desired protein is just as likely statistically to fall together on day 1 or any other day as after 10^{90} years, but that's not much of a likelihood). There are nowhere near enough atoms in the universe to construct all the possible proteins, nor has there been remotely enough time. Furthermore, life is thousands of times more complex than one single molecule of one protein: it requires thousands of copies of thousands of different proteins, all in the same cell at the same time. Life, says the calculation, could never have come into existence by chance.

The same story is sometimes set out as the Shakespeare allegory. How many billions of years would it take how many billions of monkeys tapping randomly on keyboards before the complete works of Shakespeare popped out by chance? The calculation would keep an unemployed mathematician off the streets for less than a moment ($27^{5\,000\,000}$ taps – they're not getting away without the spaces). The line 'I am a feather for each wind that blows' (*The Winter's Tale*) should appear during the next 10^{39} winters with a billion monkeys tapping once a second; 'Mine eyes smell onions; I shall weep anon' might live up to *All's Well that Ends Well*, but it's definitely not worth waiting to find out. The calculations are entirely reasonable, but the implied hypotheses are not at all reasonable because nobody ever suggested that the works of Shakespeare originated in a box of monkeys, or that haemoglobin was required to fall together.

Does the vanishingly slim chance of any known protein originating by chance disprove that life originated through spontaneous chemistry? No. It does nothing of the sort. It fails because it is self-denying. It sets up a phony prophecy that is well acknowledged to be impossible, then laboriously proves it to be impossible. No valid conclusion follows from the exercise, just as proving that the floor is flat does not mean that the Earth is flat, or proving that iron sinks does not mean that iron ships must sink, both of which were once respectable extrapolations. Since the spontaneous protein argument is irrelevant, there isn't any great point in listing all the reasons why, but here are just a few.

For a start, proteins are here and life is here, so what's the point of arguing that they are impossible? Whether their origin was natural or supernatural is a different question and is for the reader to decide.

Then there's the totally spurious constraint of trying to calculate how long one particular protein of specified amino acid sequence would have taken to fall together. Of course it would 'never' happen, we know that without the need for laborious disproof (but have disproved it anyway). The supreme exercise in futility is to try to prove that something has no chance of happening when it already has. There is every justification for believing that we have zero hope of scooping the lottery. By extension, no one will ever win it, yet last week someone won it (and the week before ...). The point always missed in the weary proof that perfected proteins did not spring into existence by blind chance is that no scientific hypothesis claims that they did.

The key distinction lies in the word 'perfected'. Proteins that did first appear would have had random structures and a variety of extremely feeble catalytic activities. A modern enzyme is not the one and only shape out of millions of possibilities that could have done a particular job, it's the consequence of a starting point. Once a few random proteins had been adopted by emerging life, the die was cast and refinement was a better option than starting again. The modern enzyme works perfectly only by definition, being the best available. How selection and refinement happened will be discussed shortly.

This chapter has shown that chemical interactions not only can happen, but must happen. The variety of chemical compounds can only increase and their dispersal on every scale from microscopic to universal is an inevitable consequence of the laws of physics. The catalogue of molecular species utilised and essential to life is large, but not astronomical, and even the paucity of genes in humans has come as something of a shock – fewer than 30 000 in contrast to the 150 000 or 100 000 confidently predicted a few years ago. Life is chemically complex, certainly, but not impossibly so – obviously. We have seen how some of the building blocks of cells could have grown around the central need for both nucleic acids and catalytic proteins, and how membranes might have appeared. The question to consider next is how the jump was made from the component parts to the finished prototype, the first reproducible cell. From this point on we at least have a model in the form of the simplest known independent cells, bacteria. What clues do these offer about how they were first put together?

NON-EVENT

The moment life did not come into existence

How can the recurring paradox be resolved: that life could never have operated with less than a certain minimum complexity, whereas that degree of complexity could never have been achieved suddenly by chance? Another conundrum can be dragged into the same debate: did intricate proteins with crucial catalytic activity appear before they were required, which seems illogical, or later, in which case how did life manage without them? To put the debate into a still larger envelope, if life did not exist in the beginning and does now, what event threw the switch from the prebiotic condition to the state we call life? What happened to mark that special instant?

Nothing. There was no special event.

Logically, if life could not have settled into its most primitive cellular phase without a reasonably comprehensive catalogue of proteins and enzymes, these must have existed by that first cellular stage; not necessarily the exquisitely tuned enzymes of today, but passable versions. If primitive life needed metabolites to prime the catalytic cycles, they must have been available. Life can never have been exempt from the laws of thermodynamics, so an energy source and the means to capture that energy must have been available (the first law). Wherever the definition is set for the minimum of complexity worthy of the name life, that entire kitset of reactions and subsystems must have accumulated in the primordial soup.

If this smells unlikely, recall two things. First, the available timescale was immense – that of the universe. The chemical components of life are exceedingly ancient. Second, a truly primordial soup, before the existence of life, could grow ever richer. If life did not exist, the planet was by definition sterile and there were no bugs to spoil the soup. It would last for all time except for chemical interactions that actually extended its diversity. The soup simply grew richer and richer.

The stage was reached where one single life form occupied the entire primaeval lake, all the oceans; the planet (any planet) was one life. The set of reactions was so diffuse as to be unrecognisable as life. Not until those reactions were partitioned into smaller and ultimately microscopic volumes did they reach the kind of velocities that we subjectively associate with living processes.

Pre-Cellular Life

The pre-cellular phase is the era in which life first came into existence and must have been the most difficult and longest phase. It is the era in which life appeared spontaneously out of nothing more than simple chemicals and energy. Such a concept is impossible for many non-scientists (and some scientists) to accept. Life undeniably descends from life. It is all too easy to extrapolate that concept over the bar between science and metaphysics and conclude that life can only ever have descended from life. Such a purported answer is so vacuous that it cannot be the response to any scientific question. The arguments of science lead in a different direction. Knowledge of the source of the universe is not necessary in order to make deductions about the origin of life.

Logically, the universe came into existence before life, therefore the early universe was lifeless. The only plausible deduction is that life emerged spontaneously some time between then and now. The simplest cell is highly complex. Highly complex life could not have appeared suddenly and therefore appeared gradually. Since everything that happens in life is without exception chemistry (unless one wants to play with words and argue that chemistry is made up of different sciences), the only conclusion is that the appearance of life was the culmination of spontaneous chemistry becoming increasingly complex, ordered and fast. As virtually all reactions of life are catalysed by enzymes, clearly the emergence of catalytic mechanisms was pivotal in speeding up life's reactions to the high-speed variety now recognised. Logically, catalysts accelerated reactions that were already possible, because that is what catalysts do.

It stands to reason that living cells could not have come into existence before the multitude of chemical reactions they initially contained, therefore those reactions appeared in primordial soup before cells developed. It is no use having the reactions of life locked up inside a cell unless a mechanism exists for the cell to replicate its genes and divide into two, otherwise life would come to a juddering halt. However, gene replication and cell division consist of an intricately controlled set of events catalysed by a multitude of enzymes. Inescapably, the reactions of life reached a complex stage in the primordial soup and on surfaces before cell membranes were invented, but would still have been very slow. The cell itself must have been a relatively late innovation in the evolutionary progression, but was probably the most crucial in allowing the rate of life's activities to increase by several orders of magnitude.

Pre-cellular life very likely centred around a chemical partnership between a simple protein-synthesising predecessor of the ribosome (a haplosome) and a reservoir of RNA priming the reaction. The answer to which came first, nucleic acids or proteins, must be that life cannot function without either and so both were required together. The popular idea of a primordial stage of life based on RNA (the 'RNA world') cannot mean that life ever consisted solely of RNA, because there is no known manifestation of life that is not essentially protein. If the idea is that RNA preceded DNA as an information store, that's different and easily acceptable, but the ultimate expression of that information had still to be in the form of protein and the products of protein activity.

A reaction retained to this day could be a significant model of the primordial bridge between the nucleic acid and protein dimensions of life. That reaction is the covalent combination of an amino acid with a nucleotide at the tip of transfer RNA. Many other forms of loose interaction between proteins and nucleic acids or nucleotides are found in nature, but are held together with hydrogen bonds, charges or hydrophobic interactions, all of which are weak. They are not true covalent chemical bonds. The transfer RNA–amino acid linkage shows that such a strong covalent bond can form under the conditions of life and provides a starting point for exploring how life could have built on a nucleotide–amino acid co-operation.

EARLY CELLS

The first cell capable of self-duplication marked the end of the pre-cellular phase of life and the beginning of the cellular phase, which could not have been achieved until the basic unit of pre-cellular life was capable of replicating in its entirety, because clearly the daughter cells needed to be complete. This supports the vision of pre-cellular life centred around a command particle – the ribosome, or primitive haplosome – carrying all the catalysts and nucleic acid information stores that could be replicated as a whole, building itself by drawing on a resource of smaller molecules in the supporting soup. Some of the required small molecules might be few and far between, but that simply introduced a time factor as the particle waited for the next collision of a desired molecule; no one set a time limit and life is very old. The absence of a cell wall was initially a positive advantage.

Eventually life was able to catalyse the synthesis of some of the nutrients required, such as amino acids and nucleotides, instead of depending on their sparse occurrence in the primordial soup. For as long as the life particle depended on receiving all its essential molecules from the environment, a cell wall would have been a real hindrance. On the other hand, once the simple system of life started to make its own metabolites, they would have diffused away rapidly and been lost unless they were retained by a surrounding membrane. At that crucial point it became advantageous for the system to be enclosed within a rudimentary cell wall.

As we have seen, the structure of a cell wall is itself highly complex since it must be neither fully permeable, which would be pointless, nor fully impermeable, thereby shutting off the cell from its food supplies. A membrane has to be partially permeable and indeed a modern membrane has selective permeability, regulating which chemicals are allowed in and out of the cell, when and in what quantities (Chapter 5). Such complexity is achieved by a partnership between lipid (fatty) chemicals that give the membrane its impermeability, and specialised proteins that in effect punch special holes in the membrane to allow particular chemicals to pass through, sometimes one way, sometimes both ways.

Modern cell membranes originate from within the cell itself. Enzymes catalyse the synthesis of a battery of lipid molecules, each having a tadpole shape with a head attracted to water and a tail that shuns water. A membrane consists of these tadpoles packed together shoulder to shoulder forming a sheet, then two such sheets are stuck together as mirror images, heads on the outside and tails together on the inside. This arrangement, known as a bilayer, gives the membrane (not the cell) a dry interior region filled with tails, with water on either side in contact with the heads. The natural chemical preference of the lipid molecules for this arrangement ensures that the membrane self-assembles if the components are available. The result is reminiscent of two strips of self-adhesive tape with their sticky surfaces stuck together, shiny sides out. If you can still get hold of the two ends, you can just about pull the strips apart – and some of the research on membrane structures has done just that.

The impermeable component of the membrane is the fatty stuff. The selective permeability is provided by an array of specialised proteins incorporating water-resistant regions that settle comfortably within the dry lipid interior of the membrane. The proteins are usually rather larger than the thickness of the membrane and the part that inevitably protrudes through the surface (often both surfaces) has a composition that is happy in water together with properties crucial to its special function, which may be to transport molecules, detect chemical signals or anchor cells together. By being equipped with predominantly hydrophobic amino acids in the middle and hydrophilic ones at the ends, the protein orients itself naturally as it floats in the membrane.

At one time life was presumably non-cellular and membraneless, but nowadays the presence of an outer membrane (and a variety of internal membranes) is diagnostic of a living cell. A plausible origin of membranes flows from the concept of a haplosomal particle increasing in complexity as the focal point of protein synthesis. Many of those proteins would be enzymes, therefore the same particle could provide the catalytic surface for fatty acid chains to increase in length, leading ultimately to the 16-carbon and 18-carbon lipids typical of membranes. Lipids of this length are not very soluble in water owing to their long water-resistant tails of hydrocarbon, so the natural destination for the newly synthesised fatty acids is to remain stuck to the particle. Lipids do in fact tend to stick to proteins, which is how they are carried around in our bloodstream without clogging up arteries. Newly synthesised lipids can be envisaged hanging around the proteins that catalysed them, until they are large enough to detach as sheets of double-sided membrane and ultimately envelope the rudimentary ribosome. Cellular life is close to having created itself, but as we see next, modern cell membranes are vastly more intricate.

THE PROKARYOTIC PHASE

Primordial cells surrounded by primitive rudimentary membranes are a very long way in concept from even the simplest autonomous cells observable today, the prokaryotes or bacteria. The structure of bacteria nevertheless shows how far such primitive cells had to develop before they could stand comparison with modern bacteria. Or looked at from the opposite perspective, it shows the richness of component parts that had to be available to early cells.

Modern bacteria, for their size, are complex almost beyond words. Certainly they are not the mere specks of cytoplasm surrounded by a simple membrane that they were thought to be when bacteria were first discovered. A powerful armory of techniques including high-powered electron microscopy, molecular sequencing and the X-ray crystallography of proteins has shown bacteria to be so sophisticated that, returning to a recurring theme, their biochemistry is not a lot less complex than yours and mine – just slightly different.

Bacteria have stunningly complex cell walls incorporating amino acids and a network of linked sugar-like molecules, an integrated structure reminiscent of reinforced concrete. Amino acids are notable for being generated in primordial soup experiments and sugars for being among the earliest products of photosynthesis, so the availability of these components should never have been a problem. Bacterial wall surfaces are often equipped with infinitesimal hair-like flagella, providing a propulsion system. Flagella actually spin to propel the organism in the manner of a ship's propeller, using a fantastic molecule-sized electric motor powered by protons (hydrogen ions having a positive charge, H^+) rather than electrons (negatively charged).

The process is closely comparable with the ATP synthase enzyme that uses a stream of H^+ to generate ATP, as described in Chapter 5. Flagella motors and the ATP synthase generator have joined the eye and the feathered wing as adored examples of how evolution cannot possibly have happened (except that it did). (The bacterium *Synechococcus* swims perfectly well without flagella, proving that they are not that essential and there's still a lot to learn about bacterial locomotion.)

All this exists in a complete living organism no more than a micrometre long – a thousandth of a millimetre! And this little lot, of which this description barely scratches the surface, with hundreds of specialised enzymes and a set of genes so long that its DNA unfolded may measure 1000 times the length of the cell itself, is so incredibly tuned to perfection that some species can reproduce their complete selves in 20 minutes. Small is not simple.

Clearly, the bacterial cell is far more than a structure, it is a process ultimately under the control of its genes that leads to the production of all the correct parts in the correct quantities, and their assembly in the right sequence in the right places for the cell to be able to divide into two. It is known that the supply of many key enzymes in bacteria is regulated in response to chemical signals that control specific genes or messengers. Perpetual growth and duplication (bacteria, by dividing, are technically immortal) is an achievement so conceptually vast that the later diversification of life into the different forms found on Earth is trivial by comparison.

In no way could primordial cellular life have been that sophisticated or had the capacity to reproduce itself in 20 minutes. Today's bacteria are not the original life on Earth and are far from simple. They have undergone 3 or 4 billion years of refinement and in that sense are considerably more tuned than supposedly advanced life such as ourselves. In a head-on battle, a one-picogram (10^{-12} gram) bacterial cell and its offspring can wipe out a human within days. The bacterial species alive today are doubtless descended from primordial forms of unicellular life, but have developed and changed out of all recognition.

Further illustrating their sophistication, organisms as minute as bacteria nevertheless may employ sexual reproduction, meaning in biochemical terms the fusion of two cells to combine two equivalent sets of genes in the fusion cell (zygote). Some bacteria reproduce sexually, some do not. Some have the option, which doesn't mean that they make conscious decisions but that they respond to the chemical and physical environment. Primitive lipid membranes, being essentially self-assembling, would have allowed early cells to fuse together readily for fertilisation and then to divide into two, but easy membrane fusion is both good and bad.

Much evolutionary refinement has evidently gone into the task of actually preventing cell membranes from fusing together, except when especially required, so for instance multicellular animals do not suddenly coalesce into a great blob and people's lips do not fuse together while they sleep. But conversely, cell to cell adhesion keeps limbs and organs where they ought to be and is indispensable for wound healing.

The massive evolutionary advantages of sexual reproduction, and the crippling disadvantages of being without it, ensured its retention. Inevitably, the stage of having many genes was preceded by having few genes. Early life was so dependent on its minimal set of genes that loss of the activity of any one gene by mutation could have been fatal to the organism. Before plant life caused the enrichment of the atmosphere with oxygen that led to the generation of an ozone shield against radiation, mutation rates were probably higher than now (although water has some blocking effect). Without sexual reproduction, a population could have been picked off one by one as different mutations ended every line of descent.

Having two complete sets of genes, one inherited from each parent, enables one gene to sustain a disabling mutation while its twin gene is unscathed. Provided that one of the genes remains active and supports life, mutation in the twin gene can be tolerated. Apart from saving the organism and the species, this provides a beautiful positive mechanism for speeding up the evolutionary process. Just occasionally, a mutation in one of an identical pair of genes would enhance the activity of the enzyme it coded for or change its specificity a little, enabling it to accept a slightly different substrate or catalyse a slightly different reaction. The mutated gene has then become a novel gene while its partner has retained its original function. Some genes exist in multiple copies, which have usually become slightly different versions although still doing a good job, so that each parent contributes maybe two or three versions and the fertilised egg cell contains four or six. Hereditary diseases, including human conditions such as the thalassaemia blood disorders, can differ in severity depending on the total number of good and defective genes inherited.

Sexual reproduction enables the growing collection of genes to spread sideways throughout the population. The simple device of having a duplicate gene has converted the consequence of mutation from being probably lethal into the foundation of evolutionary adventure. The time required for a new protein or enzyme to come into existence in the pre-life phase could have been almost an eternity, but the modification of a successful enzyme to perform a slightly different reaction and thereby create a new activity, without disabling the old, could occur in an evolutionary blink. It can now be done in the laboratory and put back into the species.

Taking a global view, sexual reproduction enables the gene pool of the species to be distributed across a wide and advancing environmental spectrum, while the spare set of genes enables experimentation with a low risk of lethal consequence. At every reproductive cycle the opportunity occurs for individuals to exchange experiences and benefit from each other's gains, steadily extending the strain's environmental territory, tolerating different temperatures, salinities, diets.

The benefits of gene duplication are overwhelmingly advantageous. On the principle that complexity was preceded by something simpler, gene duplication by sexual reproduction was preceded by gene duplication without sexual reproduction. Logically, within the earliest cells gene duplication happened before the formation of autonomous daughter cells around each duplicate genome. It could not have happened the other way around – cell division before gene division – or one of the daughter cells would have been left without a genome, a doomed blob of cytoplasm. Tracing back further, the stimulus for gene duplication was the readily replicable nature of the chemical structure of DNA, more so than RNA, even though both can carry the same information.

Fitting slightly enigmatically between the prokaryotes and the eukaryotes are the archaea. These are unicellular organisms found living in conditions almost unbelievably hostile such as beneath 2.6 kilometres of ocean, subjected to pressures hundreds of times that of the atmosphere, and temperatures of 94 °C or more in water issuing from submarine vents, with essentially no light. Others live in highly saline waters. Morphologically they appear to be bacteria and originally the group was called archaebacteria. A member of the group named *Methanococcus jannaschii*, which as its name suggests is able to synthesise methane as a metabolic end-product, has had its entire genome of well over a million bases of DNA sequenced and more than 1700 genes distinguished. The genes are an interesting mixture, with many of those concerned with metabolism being more like bacterial and those concerned with gene interpretation being more advanced (in terms of their protein products), while many of its genes are neither one nor the other. The archaea have sufficient bacterial and eukaryotic characteristics to be placed for the time being in a domain of their own, distinct from prokaryotes and eukaryotes but nevertheless derived from a common ancestor.

One thing the archaea are not, however, despite their evocative name and ability to feed on little more than inorganic materials: they are not primitive. On the contrary, their armory of several hundred recognised genes for amino acid synthesis, polysaccharide synthesis, nitrogen metabolism, electron transport, fatty acid metabolism, ribosomal structure and carbohydrate metabolism, to name but a few, projects an awesome complexity for such a minute organism. The same can be said about the many species of bacteria currently being discovered deep below the surface of the Earth and in other unlikely niches. Those forms of life have had the same evolutionary timespan as any other and a spartan existence does not equate with biological simplicity; it is equally likely to mean the opposite, a highly developed ability to survive on very little. Species adapted to extreme environments or an anaerobic metabolism are red herrings in the search for clues about the structure of primordial life.

PSEUDO-COMPLEXITY

The fundamental chemistry of life was settled at an extremely early stage that marked the end of the origin of life and the beginning of its evolution. Fundamental biochemistry has changed little in at least 4 billion years and probably much longer, but a subtle distinction is that the enzymes responsible have indeed evolved enormously since the first versions appeared. An enzyme can do the same job in a variety of species and yet look rather different in sequence and structure, but the differences are essentially neutral. They don't matter greatly provided that the metabolic pathways continue to perform.

The huge apparent increase in complexity from prokaryotic to eukaryotic cells is an illusion (as a reminder, prokaryotes are bacteria, eukaryotes are 'higher' than bacteria including amoebae, yeasts, multicellular animals and plants). The eukaryotic cell is a symbiosis, an annexation of primordial cell types that have been enslaved to become specialised organelles such as mitochondria and chloroplasts. These captured bacteria have degenerated in many ways compared with their previous free existence and have sacrificed their independence. The eukaryote is not really much further forward than the prokaryote, it just looks that way while running along similar biochemical lines. Multicellular organisms may be built of large numbers of eukaryotic cells, there are about 10^{12} (a thousand billion) in the human, but the individual cells have surprisingly little to show in the way of true innovation compared with the simple one-celled eukaryote.

So-called advanced, multicellular life is relatively recent, occupying perhaps the past 600–700 million years (little more than 400 million years on dry land). This is less than 20 per cent of life's time on Earth and 5 per cent of the age of the universe. Being multicellular is also unnecessary. Life does not of course see a necessity in itself, but if reproduction and spreading are our criteria of the success of life, then the entire living body in excess of the reproducing cells can be considered superfluous. No one expressed it better than Samuel Butler: 'A hen is only an egg's way of making another egg' (*Life and Habit*, VIII). From a universal perspective of life, multicellular bodies are a strange and inefficient diversion.

One of the most enigmatic processes in biology is the way a single fertilised egg cell multiplies into an organism as complex as a human with hundreds of different types of specialised cells, all in the right place and continuing to do the right thing at the right time for 50 or 100 years. The process relies on a succession of cell divisions in which each daughter cell may be different from its parent. This is exactly the opposite of what happens in single-celled life where, barring mutations, the cell is faithfully duplicated. If one type of cell is to give rise to another, it must contain the genetic information for both types and express it differentially when required, implying a higher level of control sophistication.

Liver cell, kidney cell, gastric acid-secreting cell, cochleal cell in the ear, retinal cell in the eye – each must appear in the correct place at the correct stage of embryogenesis and must co-ordinate with its neighbours to construct each functional organ. The theory is that crucial sets of genes program the hierarchy of cell divisions necessary for the embryo to take shape. They do this by directing the manufacture of specific proteins that in turn switch other genes on or off. Should an error occur in this process the consequences may be catastrophic for the embryo, because all events downstream of the faulty decision compound the disaster.

Much is made sometimes of the special molecules and cells apparently unique to higher forms of life, such as hormones and eye cells, but this edifice of superiority has been crumbling in recent years as the database of protein sequences has grown. Increasingly it is clear that life has played a perpetual game of refurbishing old gear. Many examples have been discovered where proteins unrelated in function seem to be derived from the same ancestral protein, revealed by key structural similarities and tell-tale pieces of amino acid sequence. Nature has not started again but has modified a previous model.

One of the most spectacular examples is the group of proteins known as crystallins found in the lens of the eye, that admittedly wonderful organ so enthusiastically focused on by those who see it as a prime example of irreducible complexity and the impossibility of evolution. How can so perfect a piece of anatomy as the transparent high-resolution lens of the human eye have come together by pure chance? (No scientist says it did, and the resolution is actually quite poor.) Several different crystallins, known as α, β, γ and so forth, are found in the lenses of various species, contributing the bulk of their structure.

It has been discovered that lens α-crystallin is related to another well-known class of proteins known as heat-shock proteins. The β-

and γ-crystallins are evolutionarily related to a calcium-binding protein called Protein S found in the soil bacterium *Myxococcus xanthus* (and doubtless many other bacterial species not tested). Bacteria obviously have no need for lens proteins and one theory is that an ancestral protein became Protein S in some bacteria that found it useful for binding calcium, but became other things in eukaryotes, thereby keeping the gene alive, eventually remustering as lens β- and γ-crystallins. Even more remarkably, the ε-crystallin of some birds and crocodiles (which are themselves closely related to each other, Chapter 2) turns out to be extremely similar to the common metabolic enzyme lactate dehydrogenase, but conscripted into a clearly non-enzymic role.

Many hormonal proteins of vertebrates have remarkably similar counterparts in invertebrates, for example insulin, which regulates blood glucose in vertebrates, is similar to a hormone that inhibits moulting in crustaceans. Mammalian vasopressin regulates kidney water reabsorption whereas oxytocin stimulates uterine contractions during birth, yet these two hormones have moult-inhibiting activity if given to crustaceans (a specialist review of this research is referenced in the Appendix). The logical explanation is that in some instances at least, hormones having different effects in different branches of life have a common evolutionary origin, which must have been many hundreds of millions of years ago to pre-date the divergence of invertebrates and vertebrates.

Nature's trick of handing in old assignments throws some light on how life's rich diversity of proteins and enzymes was first achieved. It supports the idea that many proteins may be adaptations of earlier models, perhaps a very few, possibly only one. This short-cut method elegantly solves a problem so great as to be virtually insoluble otherwise. There can be no concept of sight, feathers or any other facility coming into existence by demand, or of evolution being channelled to fulfil a subjective need. As discussed in Chapter 6, the spontaneous appearance of a protein to meet pre-selected specifications would take an eternity. Instead, mutated products have been led down new pathways and applied to completely new roles. The existence of genes in pairs has made this tolerable, as the old activity is retained while the new one develops in parallel.

Once again, the complexity of life is less than its diverse and developed state might superficially convey; the same spare parts simply have different brand names. Life's origin looks still more remote, though, as less true innovation since life became relatively advanced means more beforehand, and as we have seen, in cellular and biochemical terms life on Earth seems always to have been advanced. What was going on in the first two-thirds of the universe's existence when the Earth was not around? Whatever it was, could the fruits, or at least a few pips, have found their way here or, more to the point, avoided finding their way here? The next chapter looks into the possibilities.

SPREADING THE MESSAGE

Life is universal – but don't bother searching for it

An apocryphal and rather worn story tells of a historical ruler who wanted to reward one of his faithful generals, who had won all sorts of battles and saved the kingdom many times, or invented the game of chess, or something. 'What can I possibly give you to mark your retirement and our kingdom's gratitude?'

'I'm a man of modest needs,' said the general. 'And pretty philosophical. Tell you what. I'll settle for some rice.'

'Rice? Is that all?' replied the king. 'That's philosophical indeed. Have all the rice you want! How much rice would you like?'

'Well, I really don't know how much rice I'm going to need for the rest of my days,' said the general. 'But how about making a game of it?'

'Sure! What sort of game?'

'You and I have played chess many times,' said the general.

'And it was always your game!' retorted the king.

'I guess strategy has been my career,' said the general modestly, 'but this time it's your game. Let's imagine a huge chessboard out there in the desert. Put a single grain of rice on the first square. Two grains of rice on the second square. Four grains on the next square. Eight grains on the next square. Just carry on, doubling every time, until all 64 squares are covered with rice. Will you settle for that, king, old chap?'

'No worries, mate! But I'm not a complete fool, you know, and I do know what your little game is. By the time we get to the 64th square it's going to need an enormous amount of rice to cover it, very nearly a sackful. You thought I didn't know that, didn't you? But that's why I'm king and you're only a general. We've got masses of rice in the kitchens. It's agreed, if that's all you really want!' said the king, shaking the general's leathery hand firmly.

So the king told his master of household to get on with bagging up some rice for the old general, explaining how to start with a single grain and double it 63 times. 'You could weigh it out to save counting,' explained the king helpfully, 'just get my astronomer to calculate how much to weigh, he understands numbers.'

Calling in someone who understood astronomical numbers was a prophetic suggestion. Next day the master of household and the astronomer rushed up to the king in great panic. 'There's not that much rice in the land! Not in the entire world!'

'What are you babbling about?' enquired the king. 'The kitchens are loaded with the stuff. Have you looked in the barn? There are sacks of it.'

'You don't understand, king. The 64th square alone will need 200 000 000 000 tons of rice on it! Not to mention all the other squares. There isn't that much rice in the entire world.'

Indeed there isn't. Two hundred billion tons is about 10 million shiploads of rice. It would supply the entire modern world consumption for 400 years. King wipes rice pudding from face, but learns hard lesson about the incredible power of a doubling series, which has a lot to do with the spreading of life. Life is a doubling process as genes duplicate and cells divide into two. The ultimate power of that simple process of one becoming two, repeated, is often greeted with disbelief, but everyday examples are quite illustrative.

DOUBLING — AND DOUBLING

An old party trick is to hand everyone a large sheet of newspaper. Offer a prize for the first person to fold the sheet in half, then half again, 20 times. Go! (Some student friends I discussed this with explained politely that parties are not quite like that these days, but didn't elaborate.)

The prize is safe. It is unlikely in the extreme that anyone will fold a sheet of paper in half more than nine times. By then it has 512 thicknesses. That many layers of typical 0.035 mm newspaper will measure 18 mm thick — and incidentally be only 1/512th its original sheet size. By the time it had been folded the elusive 20 times it would be 35 metres thick (more than 100 feet) and smaller than a square millimetre. A sheet of newspaper would need to be folded 'only' 44 times for its thickness to stretch from here to the Moon.

Magazine crossword puzzles offer prizes of $1000 or so. Ever wondered why the clues seem so easy, but you never win? On closer inspection, most of the clues have two equally 'apt' solutions (in the words of the Terms and Conditions) and that's how the puzzle is composed. Pairs of matching answers are selected first, such as DecISION and DivISION, then a clue is carefully concocted to suit both answers: 'A parliamentary debate can end with a D- -ISION'. Is the deep ocean floor under IntENSE or ImmENSE pressure? With every ambiguous clue the number of alternative solutions doubles. A crossword with 20 equivocal answers has over a million solutions (with 30, a billion solutions), so don't be too disappointed when your certain prize is carried over to next week's jackpot.

All three tales are ways of appreciating the phenomenal power of a doubling series. The calculations are wonderfully easy if the following simple approximation is kept in mind:

Starting with 1 and doubling it 10 times gives 1024,
or approximately 1000

Table 8.1 shows that doubling something many times soon leads to astronomical numbers. We usually grasp tenfold increases better than doublings, so the table inter-converts between twofold and tenfold. The table says, if you double something a particular number of times, what's the answer in numbers everyone is familiar with (10-base)? Look at the second row of numbers to see how it works. Doubling a number 10 times is the same as multiplying by (2 raised to the power 10), i.e. 2^{10}, which is 1024, which is near enough 1000 for in-the-head calculations. And 1000 is 10^3, so to complete the reasoning, 2^{10} is about 10^3. Another way of putting it (check the table) is that every ten doublings adds three more zeros.

Doubling has been the key to life colonising the Earth, irrespective of whether it started here or arrived here. Purely for illustration, consider the bacterial species *E. coli*, a laboratory favourite. This microscopic cell divides (or multiplies!) into two cells every 20 minutes in the ideal conditions of a warm, nutritious soup. To grow from one cell to 1000 cells takes 10 generations, because 1000, or 10^3, is about 2^{10} (Table 8.1). If each of the 10 generations takes 20 minutes, that's 200 minutes or 3.3 hours. Growth to a million cells (10^6 or 2^{20}) takes 20 generations or 6.7 hours. Once the concentration of cells reaches a billion cells (10^9 or 2^{30}) per millilitre in the usual nutritious medium, growth stops because the nutrients are exhausted, among other reasons. That's still only 30 generations, or 10 hours, and a bacteriologist growing an overnight culture of *E. coli* in a test-tube expects plenty of cells next morning.

Extrapolation from a laboratory test-tube to the oceans of the Earth gives a result that defies belief until the numbers are checked, which doesn't take a moment. The Earth's oceans contain something like a billion (10^9) cubic kilometres, or 10^{24} cubic centimetres (millilitres) of water. The difference between 1 millilitre in a test-tube and all the world's oceans sounds astronomical. If all the oceans on the planet (10^{24} millilitres) were completely saturated with *E. coli* at 10^9 cells per millilitre, the total number of cells

Table 8.1 Doubling

Beginning with the quantity 1, then doubling it:

times doubles	power of 2	exact answer	answer in round figures	expressed as power of ten (number of zeros)
0	2^0	1	1	10^0
10	2^{10}	1 024	1 000	10^3
20	2^{20}	1 048 576	1 000 000	10^6
30	2^{30}	1 073 741 824	1 000 000 000	10^9
40	2^{40}	1 099 511 627 776	1 000 000 000 000	10^{12}
50	2^{50}	1 125 899 906 842 624	1 000 000 000 000 000	10^{15}
90	2^{90}	1 237 940 039 285 380 274 899 124 224	1 000 000 000 000 000 000 000 000 000	10^{27}
110	2^{110}	1 298 074 214 633 706 907 132 624 082 305 024	1 000 000 000 000 000 000 000 000 000 000 000	10^{33}

would be 10^{24+9}, which is 10^{33} cells. Starting from a single cell, growth to 10^{33} cells would take only 110 generations (Table 8.1, last line). At 20 minutes per generation for 110 generations, the Earth's oceans would be completely colonised in a mere 37 hours.

The volume of the Earth's oceans is chosen simply to illustrate the phenomenal consequence of any doubling process. It's chosen because numbers such as 10^{24} millilitres simply cannot be grasped by the human mind, whereas a vision of all the world's oceans might be imaginable to some extent. In reality the oceans are not a rich nutritious soup such as *E. coli* would approve of, are far too chilly, and some dramatic means of stirring the seas would be necessary to spread the bacteria around as their numbers increased. It's a thought experiment, but it makes important points.

Perhaps the most important point of the calculation is that regardless of how slowly a species did reproduce by doubling, it would still require only 110 generations to increase from a single cell to a staggering (if unlikely) billion cells in every millilitre of all the oceans of the entire world. To achieve a more modest 1000 cells per millilitre would merely reduce the time to 90 generations, not a huge difference (total cells = 10^{24+3}, which is 10^{27}, or 2^{90}). If the generation time instead of being 20 minutes were more in keeping with a primordial species growing in poorly nutrient seawater, say a whole year per generation, that only raises the colonisation time to 90 years. Even if primordial life were so rudimentary that it struggled for 1000 years to double its numbers just once, it would take but a trifling 90 000 years to dominate the oceans, less than a blink in the geological timescale. Regardless of the numbers chosen, the answer is the same: primitive life could have colonised the oceans of any planet extremely rapidly, thanks to the arithmetic of doubling.

EXTRATERRESTRIAL INFORMATION

At these paltry rates the spreading of life itself is nevertheless a surprisingly efficient way for information to be distributed around the universe. One of the great non-starters of science, if one checks the arithmetic instead of listening to SETI (Search for Extra-Terrestrial Intelligence) enthusiasts, is the notion of spreading meaningful information around the universe by radio or light waves. Allowing for the diminution of signal strength with the *square* of the distance as a signal expands in all directions, it is instantly obvious that only directional communication could be contemplated. Taking into account the narrowness of the beam (essentially parallel, a pencil beam), the futility of sending a message in the direction of any particular solar system that will have moved out of the way or disintegrated by the time it gets there, plus a host of other imponderables, it would take an average 100 000 000 years to be lucky enough to pick up a one-second snatch of communication from an unknown planet somewhere in the universe – provided that sender and receiver miraculously thought of using precisely the same wavelength (the calculations are in the Appendix). That's the best figure: at worst it might take 10^{19} years to strike lucky, which is a billion *times* as long as the universe has existed. All this presupposes that some planet somewhere hosted, perhaps millions of years ago, individuals intelligent enough to send messages (but daft enough not to realise that they would be received by nobody).

However approximate (or wrong, if you will) these numbers might be, they do seem to say that broadcasting the entire human DNA sequence into space as some kind of intergalactic Web site is a fool's errand. True, by transmitting the complete DNA sequence, which would be technically trivial, one could send enough information about a life-form to hope that an entity more technically advanced than ourselves could do what even we cannot do, turn it into a baby copy of the original, except that no such entity is likely to receive it. Any such thought experiment must be distinguished from the beaming-up idea of *Star Trek*. The complete DNA contains the information to reconstruct the development of a similar individual, but it contains nothing of the intellect or personality.

Another subtle problem mentioned in Chapter 3 is that DNA alone, even if available in a bottle rather than merely its sequence printed on sheets of paper or a Web site, is not sufficient for the creation of life. The duplication of a cell under the command of DNA requires a comprehensive set of cell components. It cannot begin with DNA alone. Even viruses, far smaller than bacteria, have to hijack a cell of the correct type before the DNA or RNA they bring along can be utilised.

In contrast, the spreading of ready-made life such as a spore is dramatically more efficient than transmitting the DNA sequence by radio over astronomical distances, because a single preserved cell has the potential to recreate its kind in unlimited quantities. The information does not arrive as a diffuse and garbled radio message but in a dense and intact little packet. No intelligence or technology is required to receive and decode the instructions; just add water. Panspermia, the notion of spores in space showering down on planets to start new colonies, is in fact a very old idea, dating back at least to Svante Arrhenius in 1903.

Some familiar forms of life really are able to survive in a state of extended dormancy. Bacteria such as *Bacillus* and *Clostridium* form spores able to survive desiccation and high or low temperatures, but they quickly reanimate when conditions are right. Satellite experiments have verified that bacterial spores can survive space conditions, particularly if they are mixed with protective clay or sandstone. A number of protozoa (*Amoeba*-like creatures) can form survival spores or cysts. The seeds of many plants are highly resistant to extreme drought and the baking heat of the desert. Survival cysts of many crustacean species, including the tiny water flea *Daphnia*, can last for years. *Daphnia* DNA thousands of years old has been successfully decoded. The so-called sea monkey eggs sold for amusement are actually not eggs but survival cysts of the crustacean *Artemia*; the cyst, only 0.2 mm in diameter, contains an embryo that can survive long-term vacuum desiccation but regenerates a tiny shrimp larva when placed in seawater. Tardigrade mites can survive boiling, freezing, very high pressure and vacuum. Sponges are able to enter a long-term resting stage.

Could life have propagated around the universe? A more realistic question is whether this could have avoided happening. There is not the slightest suggestion that *Clostridium* spores, water fleas or any other familiar species permeate the universe, but without doubt life can enter highly resistant states of long-term preservation. Vast amounts of the potential components of life such as the amino acid glycine and about 130 other molecules, including sugars, alcohol and large hydrocarbons, have been identified spectrally in interstellar space, and at the very least their arrival on planets would have given the synthesis of life a head start.

The universe has existed for three or four times as long as the Earth. In all that time the universe has not been static. Our own solar system is probably a reasonable model of many planetary systems. A glance at any of our airless neighbours such as the Moon, Mars or the moons of Jupiter leaves no doubt that vast amounts of interplanetary debris have rained down on them. The reason for the Earth having far fewer impact craters than many other bodies in the solar system is its atmosphere, which not only causes many meteors to burn up before hitting the surface, but also leads to the obliteration of small craters through weathering and the effects of life. Nevertheless, some pretty sizeable ancient impact craters have been found on the Earth, such as the 170-metre-deep Meteor Crater, Arizona, while more continue to be discovered from satellite images.

METEORS

The majority of small meteorites hitting the Earth's surface look very similar to garden stones and would rarely be noticed among ordinary surface soils (meteors travel through space, meteorites are the remnants that one finds). A notable exception is in Antarctica, where stones occasionally found in the compacted snow are so incongruous that they can only have fallen from the sky. Counting the snow strata reveals the year when the meteorite landed. Similarly, meteorites have been spotted in remote Australian desert sands and other places where they stand out.

All this stuff must be coming from somewhere and astronomers have ways of deducing from where. Meteor speeds can be estimated from radar measurements of their ionisation trails. Most have speeds between 11 and 73 kilometres per second (km/sec). These two figures are highly significant. The lower speed of 11 km/sec is the ballistic escape velocity from Earth, or conversely it is the impact speed reached by an object attracted to the Earth from an infinite distance. The upper speed of 73 km/sec is the closing speed of a head-on collision between an object travelling in an orbit around the Sun just slow enough to avoid shooting off into interstellar space (42.1 km/sec in the vicinity of the Earth), and the Earth travelling in its orbit in the opposite direction (20.1 km/sec), plus a further gravitational contribution as the object and the Earth approach each other. In theory, nothing in orbit around the Sun, in other words belonging to our own solar system, can hit the Earth's atmosphere at greater than 73 km/sec, and anything arriving at less than 11 km/sec has not fallen from outer space.

The real interest centres around meteors clocked at well in excess of 73 km/sec. About 1 per cent of them travel in excess of 100 km/sec, implying that they must have originated outside our own solar system. A recent international study was able to pinpoint directionally the star systems from which some of these objects most probably came (references in Appendix).

The case for the arrival of interstellar material on Earth therefore seems clear, but, still more significantly, the reverse can happen and material from our solar system can be dispatched across the universe. The evidence is to be found in the orbits of certain comets, of which Swift–Tuttle is a good example. This particular comet rounds the Sun at a velocity almost, but not quite, sufficient to overcome the Sun's gravitational attraction and disappear out of the solar system, but much of the material sublimated from its icy nucleus acquires even greater velocity in the process and therefore must exit the solar system. That material has to go somewhere.

The cloud of comets and cometary material beyond the orbit of Pluto, named after its discoverer Jan Oort, is believed to be the product of icy material ejected from the region of the Sun's giant planets, Jupiter, Saturn, Uranus and Neptune. However, a key calculation from recent research indicates that the majority of such material has not been captured in the Oort cloud and has also exited the solar system.

Clearly, the evidence is that material is both arriving in, and leaving from, the solar system. The same must be happening in billions of star systems throughout the universe. A proportion of their material is propelled into interstellar space. A certain amount of material arrives from interstellar space. Material can go anywhere, it can come from anywhere. The sum total is a significant interchange of material between the planetary systems of the universe, at least across reasonable distances.

In fact it's not essential to invoke visibly large objects such as comets for the transfer of life or its component molecules across interstellar distances. The fine dust of emerging life doubtless pervades space and is continually swept up by cruising planets. But there are positive benefits of protection in being associated with larger bodies: protection from radiation and, in the case of an icy comet, from total dehydration. As recently mentioned, bacterial spores have been shown to survive the conditions of space better if packed with inanimate material.

The discovery that high-speed meteors arrive here from several different stars makes the point that such phenomena are not uncommon. Any notion of our solar system being a closed system, quarantined from the rest of space, seems ridiculous in the light of the interstellar exchange of meteorites, comets and dust. The bombardment of planets with debris is observable fact. Putting the two scenarios together, the bombardment and its range of velocities extending from interplanetary to interstellar, there can be no doubt that our planet has been showered with interplanetary and interstellar material throughout its existence.

ASTEROIDS AND COMETS

Whatever that material may be, whether it came from comets, asteroids or other planets, it must nevertheless have had a still earlier existence, tracing back ultimately to the origin of the universe. Much of it may have been recycled and its previous records wiped, for instance if the material has been ejected from the Sun or a star and recondensed. But not all interplanetary or interstellar material has necessarily been recycled and some of it may have survived in solid form since long before the formation of our solar system. Solid planets first appeared within about a billion years after the Big Bang. It is equally likely that planets have been breaking up since then. Space must be littered with dust and fragments of planetary origin that could have any age dating back to the first-ever formation of solid planets in the earliest epoch of the universe.

The percentage of interstellar objects or particles dating that far back may be exceedingly minute, but a minute percentage of an astronomical amount of material can be a very large amount of material indeed. The 1 per cent of meteors well exceeding the solar system speed limit represent 1 per cent of millions arriving at the Earth every day. Countless billions of particles of solid matter are distributed throughout space, some of it in clouds so vast as to block the light from clusters of millions of stars. Whether interstellar or interplanetary particles could possibly carry seeds of life is another of those questions best answered by inversion. If our solar system is gathering up dust and debris from millions of sources within the range and timescale of our region of the galaxy, what is the chance of forever avoiding contamination with fragments of extraneous life?

The majority of the debris originating within our solar system is believed to come from the break-up of comets and asteroids. One possibility is that the asteroids or minor planets, generally having orbits between those of Mars and Jupiter but often ranging more widely, are the remnants of a disintegrated planet. A more favoured idea is that the asteroids never were an integrated planet but represent a capture of space debris into the orbital gap between Mars and Jupiter, where in different circumstances it might have accreted to form a planet but failed, possibly owing to the disruptive influence of Jupiter's gravity. That still doesn't answer the question, where did the stuff come from?

Asteroids can be pretty big, 1000 km or more in diameter. They can have highly irregular shapes, as does Eros, which is roughly 37 km long and 14 km wide. If asteroids were loose accretions of dust and minute particles, they would tend to be roughly spherical as a result of gravitation. But many are not, they are odd-shaped lumps indicative of a rigid, solid structure, even if coated with layers of loose material and dust. Close-up photographs of asteroid Gaspra taken in 1991 by the space probe Galileo, and of Eros taken in 2001 by the Near Earth Asteroid Rendezvous spacecraft (named Shoemaker in memory of the well-known astronomer), showed them to be rocky and covered with impact craters. A common-sense starting point in seeking the origin of large irregular lumps of rock is to suggest that they broke apart from even larger lumps of rock. A rough and very non-spherical 1000-km-wide piece of rock sounds suspiciously as if its parent was planet-like.

The majority of asteroids show no clear signs of originating in the previously melted and solidified material characteristic of planetary rocks. Asteroids often contain hydrocarbons and are believed to be the source of the predominant class of meteorites, the carbonaceous chondrites. A very small percentage of asteroids are especially interesting because they have a high-density metal content or a melted and solidified composition, suggesting that they came from planetary fragments. Once again, that small percentage of the asteroid population is a lot of asteroids.

The composition of comets is equally intriguing. As a rule, the nucleus contains ice and rock, often apparently carbonaceous. Fred Hoyle and Chandra Wickramasinghe, working with satellite observations of comet Hyakutake and to a lesser extent Halley's comet, saw what they took to be evidence for the existence of numerous minute carbonaceous granules. Whether such granules support their idea of space 'viruses' or Arrhenius's original idea of panspermia (life pervading space) is more debatable. Viruses, incidentally, are extremely sophisticated organisms that live and replicate only after first invading an intact and functional cell of a specific type, and cannot in any sense be a model of primordial life.

Where cometary material comes from, especially the rocky part of the nucleus, remains a mystery. A partial answer is that belts of comets exist, orbiting the Sun, some having orbits averaging about the same as Jupiter's distance from the Sun, with orbital periods of a few years. Another vast reservoir of comets resides farther out beyond the orbit of Neptune in a zone known as the Trans Neptunian Objects, or Kuiper Belt, after Gerard Kuiper. Most of these never come near the Sun or the Earth, but those that do have orbital periods measured in decades or centuries. Pluto and its moon may technically be members of the Kuiper belt, as may be the tentatively named Sedna, which is still more distant. Even farther out, far beyond the orbit of Pluto, is the reservoir of cometary material known as the Oort cloud mentioned previously, which may occasionally release new comets.

Recent research has established that the Kuiper Belt is so densely packed with comets that collisions are relatively frequent.

The terms 'densely packed' and 'relatively frequent' have their astronomical meanings: the 'dense' Kuiper Belt has a total mass about one-tenth that of the Earth and is almost entirely empty space, while 'relatively frequent' collisions are exceedingly rare! Nevertheless, over millions of years there has been a great deal of fragmentation of comets into smaller ones and a continual interchange of material between them. Pictures from the recent encounter of spaceprobe Stardust with comet Wild 2 showed impact craters on the nucleus. Those few comets that visit our region of the solar system are predominantly the smaller ones resulting from collisions and break-ups that have knocked them into eccentric and shorter-lived orbits.

SURVIVING RE-ENTRY

Asteroids and comets, although quite different, have in common that they may provide an invaluable staging post for the protection and transfer of interstellar materials to the surface of the Earth. The vast majority of objects hitting the atmosphere from space burn up spectacularly. That would be a sad ending for life arriving at the Earth after travelling for millions of years. On the other hand, it is indisputable that some objects, obviously from space, do survive re-entry. How they survive is secondary to the fact that they survive, but is an interesting area of speculation. Comets and asteroids could provide safe passage for a small percentage of the material entering the solar system, bringing it to Earth or other planets in a multistage let-down process.

The first part of the trick lies in the feeble gravitational attraction of comets or asteroids. The gravitational attraction between two bodies depends simply on their masses multiplied together, and decreases with the *square* of the distance between them. Since the mass of the Earth is comparatively large, about 6×10^{21} tonnes (6×10^{24} kilograms), its attraction for any object is proportionately strong. A tiny asteroid or cometary nucleus having billionths of the Earth's mass would have extremely feeble gravitational attraction. If interstellar particles arrive in the solar system with all manner of directions and velocities, at least some will strike lucky and find themselves travelling in formation with an asteroid or comet. An object gravitationally attracted to an asteroid or comet, provided that their trajectories are in concert and not in conflict, should impact it very gently.

This is not the same as saying that objects in space are weightless and collide without damage, which is not true. The damage when objects collide depends not on weight but on mass and would be exactly the same, speed for speed, in weightlessness, or on the Moon or the Earth. That's why the prospect of being wiped out by a meteorite is a realistic part of space drama. What matters is that an incoming object coincidentally matching the trajectory of a comet would be attracted to it gently (strictly speaking, they would be attracted to each other).

At least some interstellar material is likely to be captured in this way in the outer Oort cloud or by comets in the Kuiper Belt. Little is known about the structure of the Oort cloud, but it is thought to contain vast amounts of ice, dust and rock and to extend for thousands of astronomical units from the Sun (an astronomical unit is the distance from the Sun to the Earth). The word vast, as usual, has its astronomical meaning: the total amount of material in the Oort cloud is tiny and probably less than about ten Earth masses – very little in the context. Occasionally it has given birth to new comets, possibly as a result of gravitational perturbations from within or beyond the solar system. If the Oort cloud is dense enough to spawn new comets from time to time, it is dense enough to capture a small percentage of interstellar material – and that's a small proportion of a lot of material. As comets coalesce gravitationally out of the Oort cloud, captured interstellar material would be incorporated deep inside them.

A parallel mechanism can be envisaged in the Kuiper belt, where comets already exist and must inevitably scoop up a certain amount of space debris. Recent calculations of the competition for space in the plane of the Kuiper belt suggest that collisions, break-ups and redistribution of cometary material must be relatively common by astronomical standards, providing an opportunity for material adhering to comets to become incorporated deeper into the core.

Biological materials keep well at low temperatures. They are often stored in laboratory freezers at $-20\,°C$ or $-80\,°C$, or under liquid nitrogen at $-196\,°C$. Live bacteria, viruses and even mammalian eggs and sperm cells can be preserved in this way for years. For a life-form or advanced components entering the Earth's atmosphere, the benefits of re-entry while frozen inside an accretion of intensely cold ice are obvious. A thick shield of ice also protects against various forms of radiation damage in space. Cometary ice is likely to spend much of its existence just a few degrees above absolute zero ($-273\,°C$).

The critical step is the transfer of cometary material with its interstellar harvest of life's components to the surface of a planet. But could it survive the journey? One of the most spectacular astronomical events in the history of human observation was the impact of comet Shoemaker–Levy 9 with Jupiter in July 1994. The comet, as is the convention, was named after its discoverers, Eugene and Carolyn Shoemaker and David Levy. This is the same Gene Shoemaker whose name was given to the Near Earth Asteroid Rendezvous spacecraft mentioned earlier and who might with considerable understatement be described as having been one of the greatest planetary astronomers. Satellite images showed gigantic fireballs, some of them tens of thousands of kilometres in diameter, as a procession of 21 major fragments of the comet hit Jupiter's dense atmosphere at speeds of about 210 000 km/hour or 60 km/sec, accelerated by Jupiter's massive gravitation. The popular press competed to equate the power of the fireballs, some larger than the Earth, with so-many millions of hydrogen bombs.

There can be no doubt that comets hit planets, including the Earth. In 1908 a vast area of remote Siberia near the Tunguska river was devastated by what is now believed to have been a small comet (forget the ever-popular antimatter explanations or the alien spaceship with thermonuclear engines). The comet probably exploded just above the ground because it had become superheated through friction with the atmosphere, flattening about 2000 square kilometres of forest.

There's much more to a comet than the visible and spectacular Sun-lit part: it leaves a huge trail of debris marking out its historical track, which the Earth encounters from time to time. Giovanni Schiaparelli, the astronomer who noticed channels (*canali*, disastrously mistranslated as canals) on Mars, also made the inspired discovery that meteor showers appear annually as the Earth crosses the orbital tracks of known comets. The Leonids (named because they seemed to come from the direction of the constellation Leo) appear in November, the Perseids (after constellation Perseus) in the second week of August and the Lyrids (after constellation Lyra) in the third week of April. The meteor showers are more spectacular in years when the Earth passes closer to the bulk of the debris.

Visible meteor showers might loosely be attributed to the 'burning up' of meteor material through friction with the upper atmosphere. It begins to look from the Shoemaker–Levy 9 and Tunguska impacts and frequent meteor burn-ups as though contact between a planet and a comet means instant oblivion for any interstellar stowaways snuggled inside. But not all cometary material need suffer as catastrophically. It mainly depends on whether the arriving material survives the initial impact with the upper atmosphere while travelling at extremely high speed. Provided that it does, the chances of reaching the surface are good, but much depends on that high-altitude impact.

As would be predicted, and the pictures of Shoemaker–Levy 9 confirmed, a comet is likely to break up as it encounters the gravitational attraction of a planet. The combination of sudden deceleration as it hits the upper atmosphere is equivalent to a crushing blow, which is good from a survival perspective because it releases a shoal of slowed fragments with a range of sizes. In any case, a comet is not a unified item but a nucleus accompanied by an entourage of bits and pieces trailing around it.

Not all fragments in a shower of cometary bits would arrive in the same orientation. Spin a stone across the surface of a lake and it bounces a few times as it slows down before sinking (the world record is said to be 38 bounces!). The air–water interface where this happens has similarities to the interface between space and the atmosphere. An object hitting the upper atmosphere at an oblique angle, as some would, can gaining a lifting force and bounce out of the atmosphere, dropping back gravitationally and repeating the process, gradually losing much of its momentum before finally sinking through the atmosphere.

Friction with the upper atmosphere both heats the surface of an object and slows it down. Melting or vaporisation of material (ablation) is an ideal mechanism for the disposal of unwanted heat, so that the hot material comes away from the surface while the core remains cool. The critical question for any particular object, influenced by its entry speed and angle, is whether there is anything left by the time its speed is slowed to the point where its surface is no longer ablating away. The ablation idea is exploited to protect the occupants of spacecraft during re-entry. The surface of a spacecraft returning from the Moon, hitting the atmosphere at around 11 km/sec (40 000 km/hour), can reach 2500 °C, but by arranging for a protective shield to dissipate and carry away the excess heat, the astronauts are protected.

Very roughly, depending on arrival velocity and other variables, rocky objects of more than about 1 kilogram might have sufficient substance to survive descent through the upper atmosphere. Their surface becomes hot, but the larger the object, the better the centre is protected. The most vulnerable objects entering the atmosphere at modest speeds are those ranging from about a kilogram down to specks of dust around a tenth of a microgram. Particles within this range of sizes have insufficient substance to last the distance, so they burn up. Still smaller particles survive because friction with air slows them very quickly before overheating, then they gradually flutter down to Earth as fine dust. Such minute particles cannot be dismissed as irrelevant to the spreading of life's components. To place a tenth of a microgram in perspective, it is still 100 000 times the mass of a living bacterial cell, and perhaps 1 000 000 times the mass of a dried-up but viable bacterial spore.

When a surviving object reaches lower altitudes its equilibrium speed slows enormously as acceleration is counteracted by frictional drag (which increases with air density on the way down). For a sphere this equilibrium, or terminal velocity, is related to the square of object diameter (and to its density, which is less variable so has less effect). Smaller objects are more slowed by friction because they have a larger surface-to-volume ratio, whereas larger pieces fall faster, because the size effect is squared.

Objects of a few kilograms finally clonk down on the surface at typically 50–200 metres/second, trivial compared with the initial entry velocity of at least 11 000 metres/second and much too slowly to cause significant heating – in fact, a hot object would cool in the lower atmosphere (ask a parachutist). Contrary to common expectation, meteorites normally land cold. In any case, they are likely to land in water, which covers 70 per cent of the Earth's surface. Instances of them being picked up just after landing are not completely unknown. For instance, on 9 July 1996 Indian scientists reported that a 20-kilogram, 30-centimetre meteorite fell near a village in Pali, Rajastan desert, making a 60-centimetre-deep crater. When inspected after 'a couple of hours' it was 'cold to the touch'. A block of stone that size would retain heat as effectively as a domestic storage heater.

Many people (at least, the news media) expressed surprise that *Caenorhabditis elegans* nematode worms survived the fall to Earth after the tragic loss of the space shuttle Columbia and crew on 1 February 2003. These organisms were not merely bacteria, for which survival would be readily understandable, but 1000-celled creatures. Nevertheless that is pretty small, about 1 millimetre, and having obviously been protected by their canister from the devastation, their only hazard during descent was the final impact, of which their survival is a matter of record. As always, theory cannot dictate observation, despite Sir Arthur Eddington's famous edict: 'Don't believe the results of experiments until they're confirmed by theory!'

Very large objects hitting a planetary surface (or dense atmosphere) would enter another zone of non-survival for a different reason: the massive dissipation of kinetic energy (exacerbated by higher speed) would cause a major explosion by raising the temperature of the material to incandescence.

Compared with comets, asteroids are less attractive vehicles for interplanetary exchange if life is to survive the process, but they have a part to play. The likely origin of at least some asteroids as planetary fragments has been mentioned previously. Asteroids appear to be the original source of most meteorites falling to Earth. What has been said about the physics of the arrival and descent of cometary meteors applies equally to asteroids.

The interpretation of a meteorite depends on being sure it really is one. A meteorite is seen to fall and a stone is found, but are they the same? This is a question worth bearing in mind for anyone tempted to get caught up in one of the sporadic outbreaks of bidding frenzy for meteorites. Some meteorites have exceptionally good credentials, however, including a landing in January 2000. The object, with an original mass of about 220 tonnes, had been tracked and the trajectory pointed to an asteroid origin. Fragments fell on the frozen surface of Tagish Lake, Canada, where no other rocky material could cause confusion, and from which they were collected cleanly and stored at low temperature. These samples have already yielded evidence of a number of organic chemicals, as have many other meteorites of asteroid origin. In recent years the study of the Monahans, Texas meteorite that fell in 1998, and another dating back to 1923, have revealed evidence of water, or minerals formed when water is present, which in turn could indicate that asteroids and comets have been in collision.

Interplanetary and Interstellar Transport

Putting together the mechanisms discussed in the previous few pages, the only certainty about interplanetary and interstellar transport is that it cannot avoid happening – not necessarily bringing viable life-forms, but components, perhaps very advanced components, of life. Every step is technically verifiable. Asteroids clearly exist and pieces fall to Earth as meteorites, usually into the sea. However unlikely their survival might be in theory, there is absolutely no mileage in disputing what self-evidently happens. Many meteorites are of metallic or igneous (once molten) structure, logically associated with an origin in a planetary body with a reformed rock structure.

Comets exist. Their composition of ice, rock, nitrogenous and carbonaceous materials is not in dispute. Orbits can be measured with precision and some comets are undoubtedly on their final orbit of our solar system. After many orbits around the Sun, some of their material is headed off into distant space. If cometary material vanishes out of our solar system, equally we stand a chance of acquiring the same sort of stuff from elsewhere.

The visible comets, the Sun-grazing types, are short-lived. They are evaporating, degassing, disintegrating, dumping material along their orbital paths to provide spectacular fireworks as our planet cruises through their orbits and bits burn up in the atmosphere. Comets lose huge amounts of material as they round the Sun, hence their beautiful and very massive tails, millions of kilometres in length. This is important to the argument because the tail consists of material continually being lost to the comet, which therefore has a limited life before it is all gone. Calculations of loss rates and orbital periods show that spectacular comets of the type seen from Earth usually live less than one million years, a small blink in the astronomical timescale and far less than the 4600-million-year minimum age of the solar system (Chapter 1). The conclusion is pivotal: comets visiting our region of the solar system must have adopted their Sun-grazing orbits very recently, and therefore continue to be replaced from some source of supply.

What is the source of supply? As previously discussed, the Oort cloud is the prime candidate, a massive reservoir of cometary material equivalent to hundreds of billions of comets. For the most part they are stable and likely to stay there for a very long time, but occasionally, through collisions or gravitational disturbances, the Oort cloud is the source of the short-lived comets we see visiting the inner regions of the solar system.

Interstellar debris exists. The solar system must scoop up some of it. Some will be incorporated into comets or be swept up by asteroids. Comets and asteroids hit the Earth. Some of the material survives. Louis Frank advances a much-debated interpretation of evidence that tiny, 50-tonne-scale comets impact with the Earth's atmosphere at the rate of several every minute and these may be a major source of the Earth's oceans, which would otherwise have boiled away from the hot primordial planet (a reference to Frank's book is in the Appendix; the debate continues). The work of Nicolas Dauphas, on the other hand, indicates that most of Earth's water is original and comets contributed little.

Another source of meteorites hitting the Earth, surprisingly at first, is from the surfaces of Mars, the Moon and other bodies. A massive impact of an asteroid-sized body on another planet is likely to hurl vast quantities of material into space with sufficient velocity to escape from the planet. The impact would be capable of blasting surface material from the Moon or Mars to Earth, or in the opposite direction. The notorious Antarctic stone ALH84001, hailed in 1996 as showing possible evidence of Martian life, is postulated to have reached Earth by that mechanism. The evidence for past life in it is intensely disputed, however, and widely interpreted as some sort of physical-chemical artifact. The artificial structures mentioned in Chapter 4 as accounting for apparently fossilised bacteria could account for those in the Martian stone (an article about this is listed in the Appendix under Chapter 4).

Among planets without atmospheres, and of reasonably similar masses, the transfer of surface material from one to another would proceed on a fairly even footing. Surface material would be driven into space by asteroidal or cometary impacts, some of it being swept up by other planets. Material would travel to and fro by repetition of the process until the surface layers of all nearby planets contained a scattering of each other's rocks, plus some of their own that had chanced to make the trip both ways. That applies for planets of fairly similar masses; otherwise the balance of transfer is more complicated. The more massive and gravitationally attractive planets suffer the more energetic impacts, but the resulting debris is less likely to escape from orbit. The equilibrium therefore favours a shift of material from smaller to larger planets, both by increased collection and increased retention.

Atmospheres also complicate the equilibrium. The more massive the planet, the more likely it is to have an atmosphere because it is retained partly by gravitation. Both incoming and outgoing material would be slowed down or vaporised by the atmospheres of planets such as Earth or Venus, again making capture more likely than loss. If Mars at one time had an atmosphere (and life?), the interchange of surface material with Earth would have been less then than now. Rock vaporised in a planet's atmosphere does not vanish altogether, it contributes fumes and dust that eventually find their way to the surface. The continual replenishment of the atmosphere with smoke-sized particles of metal and mineral compounds would have made elements such as sodium, iron, magnesium and nickel available to life developing in the atmosphere itself.

Surface material very probably is redistributed between neighbouring planets in this way, but perversely that could make it the weakest link in the story about signs of life in Martian meteorites. The colossal sizes of some impact craters on the Moon, Mars and indeed the Earth suggest past impacts on a truly cataclysmic scale, while the evidence of many more impacts may have been eroded away. The visible saturation of planet surfaces with impact craters shows what an enormous amount of uninvited debris arrives in this way. After billions of years, the surface of Mars is quite likely stacked high with interplanetary junk. Mars seems to have lost its magnetic field around 4 billion years ago, soon after the formation of the planet, which may have led to the early loss of its atmosphere (the reasoning, nicely encapsulated by Eugenie Samuel, is referenced in the Appendix). Surface water and its dissolved gases would disappear with the atmosphere. Once the protective Martian atmosphere disappeared, the likely refuge of any life still able to survive (had it even originated) would have been underground and the unimpeded build-up of debris would eventually have buried or obliterated it. Stones supposedly blasted from Mars to Earth, even if they were, may have been acquired only recently by Mars, heaped far above its deeply buried life. They might have got there from Earth! The case for life on Mars needs more than a little special pleading.

PUNCTUATED EVOLUTION

The origin of life is of more immediate concern than its subsequent evolution but the arrival of comets and asteroids was a two-edged sword, possibly bringing to Earth some constituents of life, then later diverting the course of evolution. The massive release of energy with the impact of a large object does more than obliterate life by mere physical contact. An asteroid-sized object might reach sufficient temperature to explode above the ground, radiating lethal heat over a large area, or impact and send up a hot plume that would heat the surrounding air, killing life over a wide radius. The effects of a really large impact would extend worldwide by blanketing the atmosphere with dust and poisonous fumes that could take centuries to clear.

A glance at one of those familiar textbook charts of fossil types showing the geological periods known as (progressively older) Quaternary, Tertiary, Cretaceous, Jurassic, Triassic, Permian, Carboniferous, Devonian and so on reveals a striking near-regularity of around 30 million years. The periods themselves have both geological and biological characteristics, the link between the two being probably major climate changes. Some if not all of these periods marked major evolutionary phases when certain types of plants and animals predominated, but seem to have ended with mass extinctions coincident with, or possibly caused by, changes to the Earth's climate.

The energy release of a major impact yields the scientific dividend of a date-stamp on the occasion by resetting some of the geochemical clocks. The Lake Acraman crater in South Australia is attributed to an asteroid impact 580 million years ago, coincidentally followed by a marked diversification in the types of algae found in the local fossil record of that period. Traces of a possible comet or asteroid impact, without as yet the crater, have recently been found in 250-million-year-old rocks, coincident with the end-Permian mass extinction. The Morokweng crater in southern Africa dates to about 145 million years ago and the Chicxulub crater of Yucatán peninsula, Mexico, is about 65 million years old, these two dates being coincident with the beginning and end of the Cretaceous period of evolution, the latter date coinciding with the demise of the dinosaurs. The Chesapeake Bay crater, off Virginia, and the Popigai impact structure, in central Siberia, have both been dated to 35 million years ago, coincident with the end-phase of the Eocene era.

Or are these really coincidences?

A correlation does not prove a cause, but can be highly suggestive. A figure close to 30 million or 60 million years often matches the change from one major period to another, or at least a subdivision into 'upper' and 'lower' periods. Naturally the dates of the older periods are the least certain. A periodicity containing 30 million years is extremely interesting because it corresponds with an oscillation of the solar system above and below the plane of its galaxy, the Milky Way. John Matese and Daniel Whitmire in Louisiana have proposed that the gravitational influences of this oscillation have displaced showers of comets from the Oort cloud every 30 million years. The impact of just one large comet would predictably lead to a mass extinction.

We have seen in this chapter that the universe is anything but a myriad of island solar systems and independent planets, physically ignorant of each other, but rather a gigantic stirred system, like thermodynamics magnified to a grand scale. Continual interchange occurs through the courier system of interstellar molecules, dust, asteroids and comets, ranging up to whole galaxies in collision. The effects on life have been many and varied, conveying the component parts and possibly complete life in the form of viable spores from place to place. But interstellar and interplanetary interchange can work the other way, interfering with life's happy progression through cataclysmic impacts that change the course of evolution. The game of evolution is not pure and uncorrupt. Every now and then the table is upended and most of the pieces lost. An endgame completely impossible to predict has been made even more so!

Previous chapters have examined individual processes contributing to the origin of life. The next chapter shows how the interweaving of these numerous forces and components has led to a design for life that is certainly intricate and far from haphazard, but nevertheless driven by the very randomness of the universe.

UNINTELLIGENT DESIGN

Life's inheritance

At first glance, it might seem wonderfully pure to try to reconstruct the origin and evolution of life in the order in which events happened. Simply start at the beginning and see where it leads. The single most fundamental reason why life exists is that chemical reactions are driven by energy, which has always been available. The sum total of interconvertible energy and mass in the universe is not distributed evenly. The universe would be deadly boring if all the atoms and all the energy were perfectly mixed into a bland and homogeneous cosmic sponge. As with the dart-throwing exercise in Chapter 4, the one certainty in the universe is uncertainty; there is all the difference in the world between an average density of mass or energy and the real distribution. If it's not homogeneous and distributed evenly then it's distributed unevenly, resulting in localised concentrations of matter and energy. If any academic discipline wishes to claim proprietorial rights to the origin of life, it's not biology, or chemistry, or physics, but statistics!

HOW DID LIFE ORIGINATE?

Atoms and molecules are driven into random collisions by their kinetic or motional energy, making chemical reactions inevitable. Since carbon first became available in the early universe, the variety of synthetic compounds has been almost boundless. The ground-upwards approach to recreating life's design is to presume a lifeless planet, predict which organic chemicals would be most likely to be synthesised, expand the repertoire of likely reactions, and the track towards life is inexorable.

Philosophically life's chemistry can never be reconstructed that way, because the very first wrong conclusion will send the thought experiment off on a tangent from which it will never return to the right track. How do we know the right track? Because we happen to be sitting on it, at least 4 billion years further on, knowing exactly what kind of biochemistry to expect. The score has been decided in advance and the game is not being played straight. It would not be an objective extrapolation from predictable primitive chemistry towards unknown life, but the opposite, working from unknown primitive chemistry towards the well-understood target of modern biochemistry. To act honestly, we cannot feign ignorance of what is presently known and must funnel back from the diversity of modern life towards a focus at the point where life came into existence.

The parallel with cosmology's Big Bang is inescapable. Life began to take shape the metaphorical day (perhaps a billion years) after the Big Bang and has been literally universal in space and time ever since. The general case is always scientifically sounder than the special case. There is no reason to suppose that the formation of life was postponed until a special planet called Earth formed four or so billion years ago, in fact the idea is absurd. Look at the problem the other way round. Nothing could have stopped primitive prebiotic chemistry from happening as soon as parts of the universe were cool enough for conventional chemistry to occur, and countless billions of such sites soon existed throughout the universe. The simple cellular forms of life evidently present on the Earth within its first few hundred million years were in one sense extremely primitive, but in another sense so incredibly complex, biochemically and physiologically, that they could only have been the product of billions of years of previous development.

The components and mechanisms of life are often described as amazingly complex, extremely clever, highly ingenious, a beautiful design. The more one probes beneath the surface into the ingenious structures and mechanisms of cells, organelles, mitochondria, nucleic acids, enzymes and their intricate reaction centres, the more justifiable the expressions of wonder become. But these are subjective interpretations. The same descriptions can be applied to everything in the universe, living or not. Everything in the universe works and therefore is of good design. Every single atom works perfectly, but that does not equate with any input of ingenuity. A good design is something we can learn from, not necessarily a product of human or super-human intelligence.

At a stretch, it could be construed that intelligence did indeed motivate the design of life, if the definition of intelligence is sufficiently broad to mean a developed response to chemical or physical input that contributes to survival and reproduction. As mentioned in Chapter 6, the origin and evolution of life were random, but not haphazard. Different creatures have variously adopted environmental niches and habits by choice, which is an exercise of intelligence that, through selection and reproduction, has indeed affected their design – but not *by* design. There is no limit to how small a loop of stimulus and response could qualify to be called intelligence, and the same reasoning can be projected on to plants or bacteria. The intelligence that has shaped their design has been their own. Once life existed, it directed its own destiny to some extent.

TIME: NO PROBLEM

Time is an essential ingredient in the understanding of the origin of life and the universe, yet it is the most inaccessible ingredient. How can anyone see back 4 billion years, 12 billion or 20 billion?

Are these dates remotely near correct anyway, recalling the academic certainty with which the age of the Earth was proven to be only 100 million years until twentieth-century science took a hand and multiplied the figure by 45? Little more than a century ago, scientists solemnly calculated that the Sun did not run on coal (which would have given it 6000 years) and concluded, as summarised once again by Laing (1890), '...the only sufficient energy we know of is that of the mechanical force generated by the contraction of the Sun as it cools'. *We know of*, note. He and his colleagues did not know of thermonuclear fusion. The crucial lesson is that however clever science may consider itself to be now, a trivial century later, there is boundless scope to be proved utterly wrong in the future.

A common problem in attempting to study the origin of the universe and life is the need to take readings over an infinitesimally short slot of time – the epoch of recent science – then extrapolate the results back over the greatest period of time imaginable, the entire history of the universe. It's like putting my 1969 car (you know what I do for a living) into reverse, jumping out and expecting it to retrace its life and arrive back at the factory. For the first couple of metres it might do supremely well, but extrapolating that result to a point 200 000 kilometres farther back might provoke some mild derision.

Some scientists would offer the comfort that drawing a line from the present to year dot is not really extrapolation, which technically means extending a curve beyond its points, but is strictly an interpolation because the two ends of the line are marked: today, and the creation of the universe. Interpolation is scientifically more respectable than extrapolation because interpolation is filling in the reasonably probable gaps between points, whereas extrapolation is flying off into the blue yonder (what forms will life take a billion years in the future?). When the gap is as great, though, as between today's universe and its creation, or between today's life and its origin, the niceties of whether one is extrapolating or interpolating are lost. The essential point is that little is known about the intermediate points along the way and progress from the beginning until today is largely mystery.

Nobody has witnessed the universe expanding or evolution happening naturally because there hasn't been enough time since people started watching. Events akin to evolution have been witnessed, such as the appearance of a fertile Primula in 1912 (named *Primula kewensis* after London's Kew botanical gardens) from a normally sterile cross, but the mechanism of chromosome doubling by which this occurred is relatively minor. An essential ingredient in both the concepts of an expansion of the universe from a singularity and the expansion of life from a primordial soup, is time. In both scenarios it is impossible to collect data over any significant timescale. On the time line of the graph only one measurement has ever been made, in the form of a flash photograph representing today. All that's available is a photo finish showing two noses. From that photo you are expected to deduce the entire progress of the race with several horses falling by the wayside and one jockey landing in the lake, the life-history of the winning horse and its complete pedigree, plus a set of architectural drawings for the stud-farm and stables. And the licence number of the horse trailer.

The period of a century or so over which scientific studies have been conducted is utterly insignificant in the overall timebase of many billions of years. The universe may well be expanding at a colossal rate and every living species may well be sustaining mutations and changing its composition all the time, but the window of observation available to science is submicroscopic to the point that absolutely nothing has happened in the pin-prick of time since observations began.

Nevertheless, we try our best. If the timebase cannot be recorded in real time and played back directly, it must be deduced indirectly. Looking forward or backward in time is impossible –

but not that impossible! Although science cannot always deliver the impossible, remarkable progress can be made by a careful combination of observation and logical deduction, provided that the limitations are clearly acknowledged.

The universe exists and life exists. At one time the universe in its present form did not exist and it stands to reason that life did not exist either. The universe came into existence in the event of the Big Bang, a physical phenomenon that cannot yet be explained, but that doesn't mean that it didn't happen. The laws of quantum physics imply that such events are inherently uncertain in the same way as the moment of disintegration of one particular atom of radioactive carbon 14 could happen in five minutes' time or a million years' time, irrespective of the element's average half-life of 5730 years – the event for a single atom is completely unpredictable. The creation of the universe was a random and unpredictable event. With the benefit of hindsight the origin of the universe can be dated, but no amount of foresight (had anyone been around before the universe to possess foresight) could have predicted the event.

Similarly, with the benefit of hindsight the evolution of life can be mapped and roughly dated, but no amount of foresight could have predicted the shape and form of today's flora and fauna. The trail back to the origin of life starts in the present and depends on small clues. A clue is that life changes gradually and slowly. A clue is that mutations are very often not beneficial and may cause deficiency conditions in which an enzyme fails to do its job. A further clue is that a great many mutations are of no particular significance, one way or the other. Humans and birds have numerous proteins in common that do the same job but just happen to be different, mere variants on the same fundamental design. It is for this reason that medical science is learning how to interchange molecules such as insulin, and more recently body parts such as heart valves and organs, between animals and humans.

LIFE HAS CHANGED ...

The clear conclusion is that proteins can and do change over time. The difference between birds and humans does not lie so much in the possession of protein variants, it lies in the program genes that control the progression of a single fertilised egg cell into a bird or a human (but the program genes probably operate by expressing proteins). A fertilised egg is not a kit of parts, it is a set of instructions for putting together an animal or a plant. It is the instructions rather than the parts that have been refined over billions of years of trial and error.

No better proof of this need be offered than the extraordinary phenomenon of metamorphosis. Many multicellular organisms have juvenile stages that are spectacularly different from the adult. Butterflies live most of their lives as a totally different animal, a caterpillar, before forming a pupa and undergoing the incredible transformation into a beautiful flying insect. Butterfly and caterpillar are about as different as could be imagined, yet both grow, and transform from the one to the other, at the direction of the same set of genes.

The same applies to plants, which pass through alternating stages known as sporophyte and gametophyte under the control of the same complement of genes. This isn't immediately obvious because in a typical plant such as a rose, the sporophyte is spectacular and is the ordinary and recognisable flowering plant, whereas the gametophyte is practically invisible and buried away in the seed-producing apparatus. The sporophyte has the standard duplicate set of chromosomes, but produces spores containing only a single set, which grow to become the gametophyte plant. This in turn gives rise to sperm and egg, each of which has a single set of chromosomes and which fuse to form the embryonic sporophyte, with duplicate chromosomes, thereby completing the cycle. In liverworts and mosses the two plant stages are more nearly equally prominent, giving the appearance of two different plants alternating in succession but having the same genes.

The theme again emerges that much of the complexity of life lies not only in its composition and structure but in the programming role of genes. Bacterial genes have to control division into identical daughter cells. The genes of multicellular organisms also have to contain and regulate the information to place different types of cells where they fit into the integrated whole. For metamorphosis to be possible, the genes have to contain the information for two manifestations of a creature as utterly different as a caterpillar and a moth, then program the necessary conversion work.

To this day, the basic chemical components required to fulfil life's assembly instructions are quite simple if the chain of life is viewed as a whole. Plants require little more than carbon dioxide, water, combined nitrogen as discussed in Chapter 5 and a few minerals, plus solar energy, to construct themselves and supply the animal food chain. But of course, to do so they rely on the accumulated genetic wisdom of billions of years, transformed into the structure, enzymes and functions of every cell. The information content of those genes has a massive concealed energy cost in having taken countless generations of life and reproduction to acquire and refine. In the fashionable terminology of full-cost accounting, the value of life's gene pool is equal to all the energy expended by all of life throughout its entire history, and therefore beyond price (but any accountant would happily do the job, and probably charge for the sunlight as well).

The difference between mammals and fish, or birds and tortoises, is trivial at the biochemical level. They are the same piece of machinery contorted ('morphed' is the buzz word) into different shapes. Feathers or porcupine quills are in one sense dramatic innovations, but in another sense ho-hum, merely exploitations of the skin's basic mechanism for secreting beneficial extras such as scales, hair and claws. Scanning horizontally across the array of modern life-forms, any particular protein such as the collagen that gives strength to skin, the enzyme hexokinase that reacts with glucose or the haemoglobin in blood, is much the same from species to species. More significantly, if the scan is done vertically in time rather than horizontally across today's species, if one delves hundreds of millions of years into the history of life, the conclusion is the same: individual proteins have changed very little over time, and in any case most of the changes are likely to have been neutral and inconsequential. Admittedly it is impossible to take samples from hundreds of millions of years ago and the vertical comparison is in fact a reinterpretation of the horizontal comparison between today's species, but if more distantly related species diverged earlier, as is logical, and the fossil record is dated broadly along the right lines, then this conclusion is inescapable. It is now becoming possible to verify ancient DNA sequences, and thereby protein sequences, from fossils embedded in amber or other preservatives.

... BUT NOT A LOT

The interim conclusion, therefore, is that all of life is an adaptation of an underlying plan that works, and has been that way as far back as anyone can see. That in itself has remarkable consequences because it requires primitive life to have been biochemically comprehensive. The thousands of enzymes essential to life would not have evolved at enormous cost of information content without critical roles to play, because biochemistry is highly efficient. Conversely, the deletion of any critical enzyme is, by definition supported by observation, likely to be lethal. Life as we know it simply needs the full kit and, while one can easily point to species lacking enzymes or lacking whole metabolic pathways, inevitably they compensate in some way, such as living a parasitic existence and stealing their essentials from a more supportive species.

Comparison of protein sequences is highly informative about relationships between species, and between different proteins within species, but hits a brick wall roughly a billion years ago. Protein sequences so long diverged have almost no original components left in common – yet they may retain exactly the same shape. They remind one of the proverbial antique sword: two new blades and three new handles. Yet interestingly there is a value in making such a comparison, because some residual content of information is still there. Each new blade had to fit an existing handle and each handle had to fit an existing blade. The blades had to be made of material that would cut and the handles had to fit hands. These scraps of information are precious and tell us that the scope for change in size, shape, material and type of fitment was remarkably limited. Globin always has to fit haem and the haem in haemoglobin has to fit oxygen.

Haemoglobin is a protein that has probably changed its precise role and the fine detail of its shape during its evolutionary history because the oxygen content of the atmosphere has continually changed. Divergences between globins of distantly related species can exceed 80 per cent, which because of the flatness of the empirical correction curve (Figure 4.1) corrects to at least 300 per cent or 1.5 billion years (at 1 per cent per 5 million years). Even this figure is a minimum buried in the noise level of uncertainty because, as explained in Chapter 4, divergences of 500 or 1000 changes or indeed an infinite number, per 100 sites, would never appear greater than about 80 per cent or so. Independent evidence comes from intron locations (Chapter 4), which have been essentially stable in haemoglobins since before plants and animals diverged from each other.

What was life doing with haemoglobin 1.5 billion or so years ago? In modern large animals, haemoglobin is an essential transporter of oxygen to the depths of the tissues (with a few other roles such as assisting with carbon dioxide removal and helping fish to stay buoyant). Forms of life more than a billion years ago were not large enough to require an oxygen transporter and, more to the point, the oxygen content of the atmosphere was little or nil (Chapter 5). Taking these deductions together – the extreme age of haemoglobin, the mysterious irrelevance of an oxygen transporter to primitive life, and the presence of introns at least as old as haemoglobin itself and probably older – the persuasive conclusion is that haemoglobin is an adaptation of an earlier precursor. One of the theories about very ancient introns is that some of them date from the formation of the gene from smaller modules or mini-genes, which had the effect of creating a new protein from smaller modules of protein. It is inherently unlikely, in fact not realistically possible, that a protein as superb and indispensable as haemoglobin suddenly appeared, all ready for duty, by the chance fitting together of modules. The idea defies all credibility, although it is possible that the protein developed as a set of modules, working as a unit, expressed by mini-genes that later fused. What must have been created, if introns are indeed the vestiges of an early experiment in mixing and matching bits of protein, was a precursor protein that evolved gradually to become haemoglobin.

What precursor protein? Reducing the problem to its minimum, haemoglobin both binds and transports oxygen, so it may previously have done only the first of those jobs. When life was unicellular, oxygen transport would not have been needed but there would have been a major benefit in hoarding oxygen reversibly. The biochemical distinction between binding and transport hinges on the ease of reversibility. Ordinarily, the oxidation of iron by oxygen is irreversible under the sort of mild reactions consonant with life. Easily reversible binding of oxygen is a trick rarely found in chemical science but developed to perfection by haemoglobin. Logically, the precursor of a molecule capable of binding oxygen and releasing it easily was a molecule capable of binding oxygen and releasing it much less easily.

So why bind oxygen without transporting it? There are two good reasons, of which the second is perhaps the better.

The first, suggested many times, is that oxygen is dangerous, corrosive stuff because it goes around oxidising things that don't need to be oxidised, such as car bodies. Early life, the argument goes, needed protection from the small amounts of oxygen in the atmosphere and developed haemoglobin as a means of trapping it. Incidentally, haem-containing molecules also bind carbon monoxide, cyanide and azide (a form of nitrogen), which is why these are so poisonous. A weakness of this argument is that it's pointless merely hoarding oxygen in bound form without some way of disposing of it. In any case, the usually accepted source of atmospheric oxygen is from the activities of life itself. It's having it both ways to propose that life generated free oxygen and could not use it, but later it became essential. That leads to the second scenario.

The second and more attractive scenario minimises all these problems. Logically, the first truly recognisable cellular life trapped solar energy. Had it not obtained energy from somewhere it could not have satisfied an implicit part of the definition of life, namely the performance of activities, which are never exempt from the first law of thermodynamics and therefore require a source of energy. What we are leading up to is that a supply of potential energy is one thing, in this case hydrogen-rich chemicals, but utilising the energy is another. Before the atmosphere was oxidative, a store of oxygen to burn the hydrogen would have been desirable (Chapter 2). To some extent a supply of energy could have been obtained from energy-rich organic chemicals in the naturally synthesised primordial soup, but of course the source of the energy to make these was principally solar, circling back towards the idea that this was the origin of life. The dependence of life on the Sun as its primary energy input is so compelling philosophically that an early development of the means for harvesting that energy seems inescapable. Such a means is chlorophyll and water.

WHEN THE SUN GOES DOWN

Haemoglobin, ancient though it is, provided oxygen to burn hydrogen-rich molecules. These in turn were manufactured by incorporation of hydrogen into them, supported by chlorophyll and diametrically opposite to the oxidation supported by haemoglobin. Chlorophyll and haemoglobin have a lot in common. Could they have a common ancestor, far older than either?

Chlorophyll is analogous to haemoglobin in being a porphyrin protein. It's another extremely complex protein-plus that could never have come into existence in a flash of evolutionary ingenuity and therefore is by no means representative of the very earliest stages in the development of life. Water can be split into hydrogen and oxygen by the application of energy such as an electric current. The chlorophyll photosystem achieves the same splitting by means of solar energy. But why does chlorophyll – or plant life – do this?

The objective answer is mainly to utilise the hydrogen so released, in combination with carbon dioxide, for building carbohydrates, in other words for building the structures and chemicals of life. However, from a subjective point of view there is absolutely no point in creating life that doesn't do anything, in creating structures that have no function. Life is not recognised and defined as such unless it is endowed with activity. This is not to fall into the trap of assuming that life had a predestined goal or purpose; there is no rule saying that life had to form at all or had to perform in any particular way. It is a retrospective conclusion, a matter of definition that an essential attribute of life is that it lives rather than merely exists. Beautiful rock formations and crystals may well contain an input of natural energy, for example by solar evaporation of oceans, but are easily distinguished from life because they don't do anything and make no responses.

Carbohydrates built from water and carbon dioxide enabled life not only to exist but to do things, otherwise the product would have been a lifeless structure. The direct involvement of solar energy in organic synthesis stops very shortly after the splitting of water and the transfer of hydrogen to the temporary custody of a carrier molecule known as NADP (related to nucleic acids). The intervention of NADP in so fundamental a catalytic reaction supports the concept of nucleotide–protein interactions being primordial (Chapter 7). The stages that follow in the construction of plants do not need sunlight directly but happen with energy already captured in the first stage, through temporary intermediaries such as ATP (another nucleotide; Chapter 2).

Photosynthesis is essentially a process of chemical reduction, the building of hydrogen-rich, energy-rich molecules. The development of hydrogen-based synthesis and the requirement for oxidative energy were so interwoven, they had to develop simultaneously. They were flip sides of the same reaction, in fact the one cannot happen without the other, reduction of one chemical being achieved by the oxidation of another. Water is made when hydrogen reduces oxygen and simultaneously oxygen oxidises hydrogen. When primordial life acquired the ability to split water, the hydrogen went one way while the oxygen went another way. Life is efficient. If free oxygen released by the splitting of water is an answer looking for a problem, then life's need to derive energy by the oxidation of food stores is the problem waiting for that answer. Nascent oxygen, it can be surmised, did not go to waste but was immediately harnessed as the source of oxidative energy.

How? Once again, the answer depends on how far back one tries to look, but it seems axiomatic that by the time chlorophyll existed, oxygen harvesting existed. The mechanism for the utilisation of oxygen nowadays centres on the enzyme cytochrome oxidase (the terminal cytochrome). The conceptual difficulty in incorporating modern cytochrome oxidase into the primordial story is that it does not operate in isolation but as part of a wonderfully complex set of mechanisms for generating ATP by means of the equivalent of an electric current, based on a circuit of flowing protons. As we saw in Chapter 5, the process is known as oxidative phosphorylation and is performed by the electron transport chain. That mechanism in turn requires the existence of an intact cell or an intact mitochondrion so as to maintain a proton difference across the membrane. Cytochrome oxidase catalyses the oxidation of hydrogen to water. Chlorophyll is a fairly similar molecule doing the reverse. Merging the two, chlorophyll and cytochrome oxidase logically originated as a single molecule having the catalytic ability to split water, using solar energy. The reactive site, by yielding both oxygen and hydrogen, would have been both oxidative and reductive, particularly since the released oxygen would have a fleeting existence in the form of the more reactive single atoms (nascent oxygen, O) before these paired to form the usual O_2 molecules. Later, this single protein (probably utilising metal atoms and porphyrin) diverged into two as the two processes drew apart on different molecules, chlorophyll specialising in photosynthesis and cytochrome specialising in oxidative processes.

Oxidation does not necessarily involve oxygen. Chemically, the combined process of oxidation and reduction is a transfer of electrons from a donor atom, which is thereby oxidised, to another atom, which becomes reduced. Hydrogen is a reducing agent because it consists of a proton and an electron, and the latter can reduce another atom. Some of the speculative models of early life focus on the fact that in modern sulphur bacteria, hydrogen sulphide (H_2S) is split into hydrogen and sulphur, rather than H_2O being split into hydrogen and oxygen. Such a reaction has been suggested as compatible with an early anaerobic (anoxic) atmosphere. Of course such a mechanism is possible since it exists in a few species, and may well have been present in some primordial species, but the precipitation of solid insoluble sulphur has none of the advantages of reactive soluble oxygen, which can readily take part in oxidation reactions. The proven existence of the H_2S splitting reaction does nothing to preclude the existence of a primordial H_2O splitting reaction.

Separation of hydrogen's reducing power from oxygen's oxidative activity, by having both chlorophyll and cytochrome, would have been highly beneficial. The building and breakdown of molecules were becoming separate pathways, but the snag was that while the hydrogen could be stored by incorporation into organic chemicals, the oxygen was easily lost. The two sets of pathways would have needed to happen together, because only at the instant when water was being split would there have been any oxygen to drive oxidative processes. When the Sun went down, photosynthesis would stop and so would life. That is more drastic than it might sound, because life that has stopped living is dying and decaying. Remaining at rest is not a matter of the processes of life and the cell coming to a halt, far from it; without constant housekeeping activities to maintain ion levels, keep degradative enzymes under control and do a host of other jobs, the cell would quickly destroy itself by leakage and self-digestion. The next obvious leap forward after separating the pathways of oxidation and reduction would have been to separate their timing, enabling oxidation to be switched on when required, particularly overnight. Of course, life didn't know that or aim towards it, but capitalised on the opportunities as they arose.

That leads straight back to haemoglobin. An essential for life to get started was an oxygen storage buffer (strictly, a myoglobin; Chapter 5). Oxygen reacts rapidly with cytochrome oxidase but that's also the problem – it has no option but to react. The modern cytochrome chain is not pushed along by oxygen molecules reacting with cytochrome oxidase but the reverse: it responds to a demand for oxidative energy by drawing from the oxygen pool. In our present oxygen-rich atmosphere, the supply of oxygen to aerobic life is effectively unlimited and instantaneous, though it may need to be transported into the depths of the tissues by a circulatory system. Today it is there, but in the anaerobic era virtually the only oxygen available was generated by the cell itself, by its own photosynthesis.

A primordial oxygen-binding molecule would have had two indispensable roles. It would have captured some of the oxygen molecules otherwise escaping from the cell, and introduced an all-important time shift between oxygen generation and oxygen usage. In simple terms, it would have allowed life to continue after dark. It only requires the divergence of an oxygen-reacting molecule such as cytochrome oxidase into two variants, one performing water synthesis but the other holding oxygen reversibly (a haemoglobin or myoglobin), and life has captured the natural invention of oxygen storage.

LIFE EXISTS

To say that life came into existence and has flourished ever since is tautological. There is of course the theoretical possibility that life has come into existence on Earth many times, except that there is not a scrap of evidence to support it. Either way, an Earth-centred origin of life is an idea that can't win. If life originated easily and had multiple starts, the implication is that it should have done so in multiple locations throughout the universe. On the other hand, if the origin was exceedingly difficult, the timescale provided on the early Earth was inadequate.

Cellular life can only have established itself one time on this Earth and is exceedingly old. As soon as life reached the cellular stage, it would have overwhelmed and consumed more rudimentary attempts that represented a ready source of useful biochemicals. The first life-system to establish itself on a planet would inevitably remain dominant by engulfing any independent system struggling to emerge later. Pre-cellular life could never co-exist with cellular life, the corollary being that cellular life annihilated non-cellular life.

The large majority of present-day life-forms base their metabolism on the utilisation of oxygen from an aerobic atmosphere for the generation of energy. The fact that Earth's early atmosphere was devoid of oxygen does not necessarily mean that early life had a very different metabolism. On the contrary, modern life provides the essential clues to the nature of primordial life by demonstrating the twin reactions of the splitting of water into hydrogen and oxygen to trap solar energy, then the recombination of hydrogen and oxygen to regenerate water and recover much of that energy (the second law of thermodynamics, which forbids perpetual motion, precludes the recovery of it all).

No molecule is more crucial to life than water. The reactions of splitting and resynthesis of water are so synonymous with life that they were surely close to its origin. If life grew outwards from the twin reactions, then which came first, splitting or synthesis? One might assume that on a planet with water, the splitting reaction came about first. But since matter in the universe is extremely sparse (no more than three atoms per cubic metre on average, most of it hydrogen), the possibility exists of a still earlier oxygen-trapping molecule that developed in space itself, garnering rare oxygen atoms and catalysing the synthesis of water.

A paradox recurring throughout this book has been that the most primitive forms of life conceivable are breathtaking in complexity. Sweepingly simple observations such as the existence of great beds of carbonaceous rock containing isotopes clearly indicative of life's activities 3.8 billion or more years ago, soon after the Earth's crust formed, do not point to early life being simple, but the opposite. An extremely advanced level of biochemical organisation preceded the formation of that organic rock.

The idea that life originated from scratch and attained the cellular stage entirely on this Earth, let alone its evolution from unicellular to human, if the age of the Earth is about 4.6 billion years, is simply cramming in too much. Something has to give. We tend to accept benchmarks dictated by the so-called harder sciences in specifying the age of the universe and the age of the Earth, then try to fit what are libellously called the softer observations around them, such as the dating of fossils and the evolutionary tree. Should we? Are morsels of information handed down from the harder sciences sacrosanct?

Look again. The believed age of the Earth has inflated continually since people started thinking about the question. It started off as a few thousand years on the basis of estimates such as Ussher's (Chapter 1), which nevertheless were thoughtfully calculated and sincerely believed. Progressively the accepted age of the Earth expanded through tens of thousands of years, to hundreds of thousands, then millions, leading eventually to an extremely strong consensus throughout much of the nineteenth century that about 100 million years was the correct age. There's all the difference between a consensus and a right answer, however. Newly discovered twentieth-century technology peremptorily multiplied that figure by a factor of 45! Is that the end of the matter? Does twenty-first-century science know everything?

The age of the universe, deduced from the combined rigours of mathematics, physics, astronomy and chemistry, is now believed to be between 10 and 20 billion years, with a consensus around 13 billion years. But this figure is as uncertain as the key Hubble constant (H_o) on which it is most often based and the still more uncertain shape of the expansion. The age could be revised upwards in the light of recent evidence that expansion may be accelerating and therefore was previously slower.

Time, however much of it there may have been, is no barrier to the idea that life's beginnings date back to soon after the creation of the universe and long before the Earth appeared. Only humans entertain a concept of time and therefore speed. Life's catalysed reactions now happen so fast that, by definition, the chemistry of pre-life was inestimably slower – although only by human standards. No possible concept of time or speed governed the formation of life. All of time was available and indeed that is what it took. No intelligence necessarily directed the process in the past and the intelligence that will direct life's future is its own.

Concealed somewhere in the history of life is evolution's own expansion constant. Life has been expanding, developing, growing, evolving at a rate of which the dimensions are not yet clear, but a constant is hidden in there somewhere. Once discovered and extrapolated to zero (always dangerous!) it will date the origin of life as Hubble's constant dates the universe. Will life itself set new limits to the age of our planet and the age of the universe?

The next chapter shows the enormity of trying to predict how life will be in the future, or what form it might now take anywhere else in the universe, reinforcing the stark reality that our own supposedly advanced form of life was anything but an intentional design. Predicting the course of life would be like walking into a library, reading the first word in a book chosen at random, then using it to predict the last word in another book chosen at random. We can marvel at the incredible diversity of life around us and muse over the seeming cleverness of how it all fits together (mainly by species killing each other), but we are observing the result of millions of switches thrown and turnings taken, all randomly – but not haphazardly.

LIFE: TO BE CONTINUED?

Life could do better, but probably won't

How shall we decide to evolve? Wings would be rather nice. Those wonderful old sculptures and paintings of angels all seem so perfectly proportioned artistically, they're almost convincing, and after all, we are said to be only a little below the angels. Sadly a little way down is a long way up and the Michelangelo-style creation, allegorical though it may have been, just wouldn't fly. Weary calculations about the depth of chest muscles necessary (a metre or so) are not the essence of the problem. The real difficultly is that vertebrates have two pairs of limbs, never three.

LIMITED OPTIONS

It is completely impossible to suggest how life may have been if it had evolved differently; the same applies to anticipating life on other planets. No one (except your stockbroker) can ever say how things would be today if history had been different. The neutral theory of evolution grants that mutations happen randomly. Yet life did not evolve randomly in the sense of haphazard; far from it. Any form of life existing today is at a point on a developmental time line. Looking back over its evolutionary history, the mutational options available at any stage were always severely limited. The species could not create evolutionary options for itself, it could only accept or reject the options presented to it by random mutations. Acceptance or rejection could mean living or dying. The majority of such mutations disappeared as unsurvivable, some very quickly; some have spread throughout the population and been retained, but that doesn't mean that they were especially beneficial – just different.

Look at it the other way: every base in all the DNA of every gene we now possess (about 3 200 000 000 base pairs and 30 000 genes in a human) is a historical mutation, an accident. Life as it exists is the consequence of vast numbers of random mutations, including major events of gene reorganisation, such as gene duplication or exon modules shuffling around (Chapter 4). Some mutations, particularly in the program genes governing the development of shape and form in an individual, have founded branches of life as distinct from each other as insects and vertebrates. Life cannot exploit mutations that never happened and we have absolutely no idea how life might have turned out if they had happened. No force attracted or guided the development of life in the direction of animals or plants, humans or vertebrates. Evolution was not pulled but was pushed, upwards and outwards like an explosion, and where the bits landed was a combination of chance and survival.

Especially survival. If conditions on Earth had been different, alternative forms of life would have been more successful and would not have led to life as it is today. An unimaginable species completely different from the human would have considered itself supreme. Of course, the conditions on Earth have almost always been different from today. Temperatures, air pressures, composition of the atmosphere, availability of oxygen, sea salinity, penetration of sunlight, competition with existing life, all have changed continually. Adaptable life was favoured. Life on other planets would not have followed the same pathways as on Earth.

The few features of the Earth's environment that have remained reasonably constant, such as gravitation and the intensity of solar radiation (even the Sun has been evolving, its radiations have increased by a few per cent and atmospheric filtration has changed), would almost certainly be different on another planet, affecting all other environmental conditions. Life on Earth is so adapted to our unique Sun–planet combination, at a unique period in its continually changing history, that life in the form familiar to us is certainly unique in the universe.

Of course life is successful, to a degree. One can scarcely stand here and say otherwise. Unsuccessful branches of life have doubtless appeared repeatedly and left little or no trace. Every form of life is successful in its own special way. Humans, with their power of thought and their capacity to consult the archive of accumulated knowledge, are intent on judging other forms of life according to specially contrived rules that rank humans as supreme. But are we? Is an oak-tree that lives for centuries on air, water, sunlight and a few minerals any less successful than the human who chops it down for firewood? The only real difference is that the oak-tree doesn't know how incredibly successful its kind is. What would a thinking oak-tree make of the activities of the human race and the 'improvements' wrought upon the world?

The narrow criterion against which humans often choose to measure the relative success of different forms of life is fondly called intelligence, but what's that? It seems to be an ill-defined combination of consciousness, stored information, data-processing capacity and a variety of subjective factors. Use of the word 'intelligence' here is for want of a better word. It applies only in the self-evident sense that a certain kind of activity, associated with the existence of a brain, is clearly present in animals but not in plants or inanimate objects, and different categories of animals perform more or less of such activity in different ways. It has nothing to do with discriminating between individuals in selected feats of mental prowess.

Speculation about intelligent extraterrestrials is an outstandingly baseless topic except for the way it reveals the human-centred perception of how life might have been. Nine times out of ten, extraterrestrials are unthinkingly depicted as pseudo-humans, with a few trivial modifications along the lines of green skin or stalk eyes. We want them to be different – but not that different. The Earth-bound designers of extraterrestrials imagine that they have concocted something as different from a human as conceivable but, leaving aside the problem of having to find employment for human movie actors, where are the limits of this imagination? Vertebrates standing upright on two legs with two arms (stun gun in right hand), a head and a brain, stereotyped males and females, and depressingly familiar values of possession and aggression. However hard writers of fiction try to concoct alternative forms of life, they stray hardly at all from the familiar human pattern.

The Human Shambles

It's not as though the human model is a particularly brilliant achievement, worth emulating. Contemplate the shortcomings of the human structure, so ill-adapted to our planet that if we react to its natural pull of gravity – fall flat on face – we can suffer broken bones or brain injuries that would have fatal consequences without medical intervention. Quite a small puncture can allow mortal loss of blood or a fatal infection.

And what an utter shambles our breathing and eating arrangements are. Land vertebrates (and some others) have a lung connected to a nose for breathing and a stomach connected to a mouth for eating, which superficially seems fairly reasonable. Unfortunately, the design is a bit twisted. The windpipe or trachea, connecting the nose to the lung, has to cross the oesophagus leading from the mouth to the stomach. The logical engineering answer would have been to keep it simple: let the two tubes run past each other, one to the left, one to the right, like water and gas pipes to a house. Nature has always shown a marked disdain for non-symmetry, however, and insisted on keeping both pipes symmetrically in the centre line of the body. As a result, they cut right through each other and are connected where they cross. The nose is connected also to the stomach and the mouth to the lungs! So every time food or drink is swallowed, a special flap (the epiglottis) has to close off the windpipe to prevent material from entering the lungs.

Failure of this clumsy valve sometimes kills people when their airways block with food or their lungs flood with liquid. The choking and spluttering reflex is an emergency attempt to clear the airways, but nature created the emergency in the first place. If the passages were separate, the airway would never block and a temporary obstruction of the oesophagus would matter little. Of course, we would need to speak out of the nose, or somewhere. It could be argued, on the other hand, that the human race owes its pre-eminence to an ability to communicate by speech, which was made possible in its familiar form by the unnecessary incorporation of the mouth and tongue into the breathing arrangements of vertebrates.

The human skeleton has a long way to go before becoming properly adapted to the upright posture. Its worst feature is the vertebral column itself. Look at any classic quadruped and see how the backbone is hung out like a chain, dipping in the centre. The quadruped backbone resembles an elegant suspension bridge with pillars at each end in the form of legs. The weights of the head at one end and the tail at the other help to pull the two ends apart while the gravitational sag of the chain draws them in – a superbly balanced design.

The quadruped backbone is under tension and being gently stretched. In the human it's the exact opposite, under compression with all the weight of the upper body heaped on it, tending to crush the vertebral bones and cartilaginous discs that would be under gentle tension in a quadruped. We help matters by lifting grand pianos or trees and then wonder why we suffer backache. The crucial joint between the human leg and hip is not exactly overengineered either, the ball head of the femur tending to snap off catastrophically if overloaded, particularly in old age. The human is trying to be a biped without heeding the lessons of the kangaroo or the dinosaur discussed in Chapter 2.

Once human behaviour is added to the equation the overall picture of the species is dismal to contemplate; impatient, aggressive, possessive, destructive of its environment, a skeletal and physiological disaster intent on preserving troublesome genes through medical intervention. No amount of eye correction or dental care will see humans born with better eyesight or longer-lasting teeth. The human race sees itself as something superior, a pre-eminent species at the peak of its achievement. It sees itself as the pinnacle of evolution, the culmination of the great formative phase of life, as though life were drawn to a purpose in the creation of the perfect human.

The reverse is more likely to be true. Seldom willingly contemplated is the possibility that the human species is transient and doomed, ultimately to be displaced by something far better. In reality, the human is a brief experiment that has occupied a thousandth part of life's existence on Earth at the very most, or considerably less depending on when the 'human' is deemed to have appeared. Far from human life being the ultimate achievement of any evolutionary process, it is a transient moment in the evolutionary timescale. A hundred million years or a billion years into the future, today's vertebrates may appear to have been an amusing blind-alley, with humans a hilarious and extremely short-lived aberration (if any record of them exists, other than a thin stratum of compressed wreckage and pollution in the geological record). We are not at the end of evolution but at the very beginning of the great new phase in which life has the power to communicate instantaneously over great distances, to record information for all time, to invent technology, to control biology and thereby the destiny of all life, the planet and eventually the entire solar system.

The interesting question is not how life has been over past aeons but how it will be in millions, billions of years into the future. No one could possibly have looked at a primordial bacterial cell and predicted humans, dinosaurs, butterflies, coconut trees or an eight-legged creature that spins a silk line out of its mouth to catch flying animals and kills them with venom. Likewise, no one has the remotest idea what unimaginable form of life will control the planet in a billion years' time. The present human form offers not the faintest clue.

Prospects

Truly, as I write this, a hang-glider just flew overhead! The obsession with flight is one of the most fabled of human dreams, depicted to saturation in mediaeval art. The Greek mythological characters Daedalus and Icarus escaped from the labyrinth by flying out on wings of wax, but Icarus flew so close to the Sun that the wax melted and he fell to Earth.

The wait for humans to evolve wings will be a long one. The problem is the elusive third pair of limbs (legs, arms, wings). Vertebrates now get just two pairs of limbs, or only one pair in the case of some flightless birds and none at all in snakes, but these families have abandoned limbs bequeathed to them by their four-legged ancestor and the remnants of the missing limbs can be spotted in the skeleton, or at least traced through evolutionary history. They are honorary quadrupeds. Probably an ancestral vertebrate had more limbs, very likely a pair associated with each vertebra or body segment, but as amphibians evolved from primitive fish only two pairs were retained. Four legs, allowing one to be lifted without falling over, are so undeniably efficient for support and movement that the setting of the number must have been programmed deeply into the vertebrate development plan.

The human's allotted two pairs of limbs have already been allocated to other functions than flying. The enhanced flexibility of the forelimb must have helped the development of manual skills in primates by allowing the arm and hand to explore almost anywhere, which the hindleg cannot. So where is a third pair of limbs, for wings, expected to come from? If the archetypal vertebrate had genes for more than two pairs of limbs, they were lost aeons ago. As we have seen, evolution does not go into reverse on that sort of scale and the lost limbs will never be recovered naturally. There is no evidence that a mutation to duplicate one of the remaining two pairs of limbs has ever happened. If it did, it was short-lived and not sufficiently advantageous to spread throughout any population.

Very rare appearances of an extra finger, for example, are not truly extra limbs but in general terms are a forking of part of an existing limb. This does happen, and in humans the extra piece is often surgically removed nowadays for conformity and convenience. Interestingly Robert Chambers, the pioneering evolutionist and publisher mentioned in Chapter 3, had six fingers, but I have seen no comment on whether this may have awakened his interest in evolution.

How, then, is the hypothetical third pair of limbs to be grown? Forward the genetic engineers ...

But wait! It's not that simple. Horrendous problems of anatomical engineering need to be thought through first.

Hindlimbs and forelimbs are not as similar as might be suggested by a first glance at an intact animal, even a quadruped on all fours. The hindlimb is articulated directly into a pelvic (hip) bone fused rigidly to the side of the vertebral column. This arrangement is ideal for transmitting the power strokes of the hindlegs into forward motion. The forelimb, if a leg, has the subtly different function of fine-tuning the locomotion, acting as a steering system, guiding the animal between obstacles and keeping the anterior part of the body off the ground. It is not the primary source of propulsion. The wing is likewise a forelimb that keeps the anterior part of the body off the ground, indeed eventually the entire body, and also provides the exquisite control needed for a bird to land on a swaying twig in a crosswind, to catch an insect in mid-air or to hover in the manner of a humming-bird in the neck of a flower. It is also the sole source of propulsion once airborne. The wing must provide both fine control and immense power. Hindlegs, forelegs (or arms) and wings are all quite differently engineered, wired and controlled for their contrasting roles.

The thought experiment of engineering arms and wings on to a creature already having hindlegs leads to instant strife. Arms and wings are both versions of forelimbs and would be in serious conflict, because they would need to be attached to the body in the same way in the same place. Forelimbs are not attached directly to the vertebral column as are hindlimbs, but suspended by a more delicate piece of engineering that forms the pectoral (shoulder) girdle. The forelimbs float freely, articulated to the shoulder blade, which is held in position by muscles and connective tissues, except for the easily broken collar-bone present in some species. In birds the bones of the pectoral girdle have extended to form a kind of cradle supporting the chest so that the bird in flight is literally lifted by its wings, but the principle of a floating rather than a rigid attachment is the same. The vertebrate forelimb has a range and delicacy of movement in three dimensions not possessed by the hindlimb, which is limited by connective tissues and muscles to a mainly backward and forward motion. You can shrug your shoulders but you can't shrug your hips.

The human wing would need to be neither forelimb nor hindlimb, but a newfangled midlimb. Somehow a unique and original design must be concocted, having a forelimb-style floating attachment with maximal flexibility and control, but positioned directly above the centre of gravity, attached to a ribcage that surrounds and supports the weight of the body as in a bird, without conflicting with the existing shoulder system. The musculature required for a human to defy gravity would be prodigious. There's infinitely more to it than designing and generating new wing bones, even if the entire set of genes could be borrowed from a bird. How would the limbs be controlled? A new and independent set of nerves and muscles would be required, needing new pathways and branches down the spinal column from a new co-ordination centre in the brain. Probably the brain and therefore the skull would need to be enlarged.

If this frivolous diversion is no more than a thought experiment, it's no less than one either. A surprising number of the component parts of the experiment are at least partially understood. Not so long ago I might have written (okay, I did write) that 'inducing vertebrates to generate a third and different pair of limbs is a thought experiment that will for ever remain exactly that. It's a notion in company with some of the classic thought experiments of science, such as ageing twins at different rates by sending one of them on a journey at almost the speed of light, or sending a starship full of hibernating crew to a distant galaxy, or recreating dinosaurs from ancient DNA sequences. In vague outline the principles are understood and instructive, but the experiments will never be done ...'

The gathering pace of gene research has left us standing. Experiments have already been done to induce arm-like characteristics into mouse hindlimbs, and leg-like features into bird wings. The genes were inserted by putting them into viruses, then using these to infect the target species. Changing the character of a limb is not quite in the league of adding completely new limbs, but not so hopelessly far behind, either.

Some of the studies have been done on insects, particularly *Drosophila*, the fruit-fly. *Drosophila* keeps company with a number of other favourite species for biological research in being worked on because it has been worked on before. In common with the laboratory rat, the bacterium *Escherichia coli* (*E. coli*), the pea-plant, the cockroach, the haemoglobin molecule and other old favourites, *Drosophila* has spawned an enormous database of knowledge into which fresh findings can be fitted very easily. This is an excellent reason to choose it for observations. *Drosophila* has taught biological science much about how animals are built of segments. The majority of animal species develop as a succession of segments from head to tail, segments that, to echo an earlier sentiment, are either very similar or completely different depending on one's point of view. Superficially it looks as though nerves, the gut and other systems running through these segments are installed independently, but in reality they are assembled piecewise, each segment contributing its fair share and then all the pieces are linked together.

A vast number of *Drosophila* mutations have been identified, many of them having profound effects, duplicating or deleting entire segments or appendages. The effect of a single mutation can be as dramatic as the duplication of a wing-bearing segment complete with an extra pair of wings. Observations of this sort lead to the realisation that, in principle, a biochemically simple genetic transformation can have far-reaching effects on the entire organisation of an advanced body plan. It is by this mechanism that the truly major divisions of life, such as insects and vertebrates, are believed to have come into existence. Many of the genes present in insects such as *Drosophila* have counterparts in vertebrates, which is not surprising if insects and vertebrates had a common ancestor. By analogy, there is no overriding reason why segment duplication should not happen in a vertebrate and introduce duplicate forelimbs, which might then specialise differently. Of course, there would be little ethical support for doing this intentionally.

Abandon the Human

Any sense of fun in upgrading the vertebrate is entirely misdirected, because there is no reason for evolution to have found its way to the vertebrate model in the first place. It might be better to start again than to try to improve on a poor start. The vertebrate pattern can be evaluated only in retrospect. It's here. It seems to work to some extent. In other ways it's a disaster, structurally and behaviourally. It's easy now to trace the evolutionary development of the human back towards the early vertebrate pattern by a combination of palaeontology and embryology, but it was certainly not a preordained evolutionary target.

This conceptual difference between looking forward and looking backward is crucial to an understanding of evolution. Looking back, the evolution of humans or any other animals or plants has an absolute certainty about it – it happened – which is easily confused with inevitability. The mere fact that early life headed down wildly different paths, leading to plants with cellulose walls, vertebrates with an internal skeleton, arthropods (insects, crabs) with an external skeleton and a number of other avenues such as jellyfish having no skeleton at all, shows that life could have headed anywhere. There was absolutely no inevitability about life reaching its present forms, nor is any pathway mapped out for its distant future. Two legs are no better nor worse than four, six, eight, a hundred or none, all arrangements are found and all of them work.

Only with the crystal clarity of hindsight can we point to the differences between humans and other major groups of animals and claim that the vertebrate pattern was best. Other hypothetical patterns than the human may have far greater potential to lead to an efficient, peaceful and constructive species but they remain hypothetical, which is why it is fruitless to project intelligence onto other familiar species. In any case, who is to say that peacefulness, constructiveness and contentment are natural virtues of life? Only humans appear to be capable of subjective reasoning and value judgements, and then only some.

Science fiction aliens, if not humanoid, are usually some sort of gigantic insect. Why? It may be another sign of the limited imagination of their inventors, not to mention the ready availability of natural specimens for the microscopes of the special effects department. Possibly one of the emotional attractions of many insects is their relatively large heads, eyes and by implication brains, features that people associate with intelligence.

Could insects assume the role of intelligent master species? The structure of their breathing or ventilation system is considered to be one of their major limitations. Insects conduct air in and out of their bodies mainly by a process of diffusion through fine tubes. This works in small insects where distances are short, because diffusion is governed by laws of physics involving time and distance. Where distances are small, diffusion time is short and the process works. It also works better when the oxygen pressure is higher and there is evidence that some exceptionally large insects existed hundreds of millions of years ago, when the air may have contained up to 35 per cent oxygen compared with the present 21 per cent (Chapter 5). However, scale up an insect to the size of a human and diffusion distances become so great that gas exchange would require too much time to happen efficiently. Therefore giant insects are impossible.

This well-rehearsed argument is in fact slightly spurious, for several reasons. To begin with, an association between size and intelligence is dubious. Are whales proportionately more intelligent than rats? Leaving aside that whales are currently in vogue and rats are out, are whales really more intelligent? Size or intelligence is not a criterion of supremacy, as evidenced by the ability of a bacterial cell having neither of these attributes to destroy a mammal 10^{17} times larger in a matter of hours. The development of intelligence and particularly consciousness in humans seems to be more a matter of enlargement of a specific part of the brain, the cerebral or neopallial cortex, rather than the whole brain. There is no knowing how large an insect brain would need to be to possess the intelligence of a human. If the insect breathing system had developed a sufficiently enhanced level of gas circulation and exchange, especially to the brain, insects might have achieved either the size or the intelligence of humans. Nevertheless, as said earlier, evolution can never chase a target and mutations that never happened cannot be selected. Alternatively, the insect brain might have developed super-computing powers within its much smaller size. Built into the human concept of intelligence is a notion of speed in data processing, yet in a different form of life with a different metabolic rate and different values the speed factor could be orders of magnitude slower or faster. Ultimate achievement has nothing to do with how long it takes.

Whether intelligent mega-insects are feasible or otherwise is not the point when speculating about how life might have been. All we can really say is that other branches of life have not, or not yet, achieved the level of so-called intellect that we choose to equate with the human. One of the few possible pointers towards an evolutionary future lies in the ability to store and apply knowledge, because that attribute has given the human species the power to change the planet, to select for or against other forms of life and, in recent years, the opportunity to change its own biology. Since other species, whatever their size, do not appear to have the ability to acquire the same sort of knowledge on anything approaching the human scale, the inescapable conclusion is that these mental faculties are biological and part of the structure of the human species, which is the consequence of its genes.

BRAIN AND BEHAVIOUR

Nature capitalises on success and the foreseeable future for mental activity is that it will be selected for rather than against.

Evolution remains an essentially neutral process ('nearly neutral'; Chapter 3). Unseen, unintentional changes have spread throughout the lineage now known as human, leading over hundreds of thousands of years to the expansion of certain mental faculties. In the prevailing environment these faculties are valued and further developments will increase the dominance of the human species until some catastrophe wipes it out, millions of years, or perhaps only thousands, in the future. But such genetic developments in brain power can never be promoted by mental gymnastics, they can only occur neutrally and be retained if they happen to be advantageous.

To see why this is so, it is necessary to dissect the nature of human mental activity. Doing this leads to some uncomfortable and depressing realisations. Superficially it might appear that human society has been improving itself over hundreds of thousands of years of civilisation. This appearance is an illusion, because society does not exist outside of living individuals. Society is a thin veneer over programmed behaviour. Rules and conventions are an infrastructure that people are not born with but born into. Evolution happens extremely slowly according to our subjective sense of fast and slow. There are four or five human generations per century and only about 100 generations have elapsed since the time of Julius Caesar, about 250 since Stonehenge. A mere 5000 generations separate us from a Stone Age person of 100 000 years ago. In any case, as discussed in Chapter 3, numbers of generations may have little bearing on the rate of evolutionary change if evolutionary clocks depend mainly on absolute time. The number of generations and the elapsed time since the dawn of civilisation, or even since the Stone Age, have been trivial in evolutionary terms.

At this point some double-talk is needed. Philosophically we need to argue both ways. On the one hand evolution is happening continually, on the other hand it happens so slowly that over the human timescale there hasn't been any! Every species is changing all the time at the level of its genes, its DNA, yet spectacular innovations are virtually never observed. Every new individual may have numerous new mutations, but few are ever adopted species-wide. Nevertheless, every generation is a small evolutionary distance from its parents, or as Charles Dickens so aptly put it, 'Every baby born into the world is a finer one than the last' (*Nicholas Nickleby*). We must all be a little different from ancestors of thousands of years ago. Indeed, we are all different from each other and it is no more than a statement of fact that members of the human species may be recognisable for their ancestry from different parts of the world.

Most mutational changes adopted are of little or no consequence, according to the nearly-neutral theory. Evolution almost always moves sideways, with leaps forward being occasionally great but exceedingly rare. Assume for illustration an average species mutation rate in terms of protein changes of around 1 per cent per 10 million years. This means that since 100 000 years ago in the Stone Age, the degree of change to be expected in the protein expressed by a typical gene would be as little as 0.01 per cent. That's just one change in every 10 000 amino acids. But proteins are typically no more than about 300 amino acids long and 0.01 per cent of a 300-amino-acid protein is only 0.03, or one-thirtieth, of an amino acid. Of course, you can't change one-thirtieth of an amino acid and it's the same as saying that in 100 000 years only one in every 30 genes has mutated sufficiently to produce a protein just one amino acid different.

The real result could differ from the average owing to the Poisson effect (Chapter 4) and would not be spread so evenly across all genes. As might have been anticipated, evidence is emerging that mutations have been selected in genes that have speeded up neuronal communication in the brain. Recent research indicates that in the Y chromosome possessed by human males, entire genes have been inactivated (or perhaps transferred to other chromosomes) at the high rate of about five genes per million years. Nevertheless, large-scale effects on genes, such as complete deletion or the transposition of large chunks of DNA sequence, are subject to the same filter of acceptance or otherwise, in other words selection, that governs point mutations.

The latest figures indicate that the human has probably fewer than 30 000 genes. If nominally 1 gene in 30 may have mutated sufficiently to change a single amino acid in the past 100 000 years, that amounts to only 1000 genes changed. Since 10 000 years ago the change would be a mere 100 genes, and since 1000 years ago a miniscule 10 genes (as amino acid changes). Taking into account that survivable mutations are more than likely to be harmless and inconsequential, these changes are so small and random that there is no particular reason to believe that we are different in any material way from someone born 1000 or 10 000 years ago.

A popular objection to this argument is that domesticated animals and cultivated plants have been bred out of all recognition compared with their wild relatives, thereby demonstrating rapid evolution, so humans have presumably evolved equally rapidly. Objectors point to poodles, weird goldfish, domestic tomatoes or garden roses and say that these are so diametrically altered from their wild relatives as to be virtually unrecognisable (and incapable of surviving in the wild). Flip through any garden catalogue and choose your new life-forms! They never existed until plant breeders started tinkering. Clearly, evolution has been extremely rapid.

It hasn't. The objection is misguided because the wonderful new varieties owe little or nothing to natural evolution. They are varieties concocted by selecting genes that already existed. The global gene pool for any species harbours numerous alternatives scattered throughout the population. The breeder who comes up with something seemingly new and sometimes bizarre has simply brought fresh combinations together. The genes themselves have scarcely mutated within hundreds of thousands of years, but there are plenty of variant genes to choose from in the pool. Sometimes the local pool starts to look a bit shallow and breeders go hunting for fresh wild stock in the hope of discovering a few new characters. Modern technology, meaning the human species, has muddied the water further by enabling the introduction of genes from 'foreign' species and artificially modified genes to create attractive transgenic varieties. Already genetically modified pet zebrafish are on sale that glow red and green in the dark. Life will never be the same again.

The argument that humans have not changed for years has to dovetail with the self-evident fact that they have changed from their pre-human ancestors. The antiquity of the human species is a contentious matter, but if we work on 1 million years for round figures, roughly one protein in three has undergone only a single amino acid change in all that time (from the calculation above – 1 in 30 over 100 000 years, so 1 in 3 over a million years, on average). How could so small a change, bearing in mind the neutral and inconsequential nature of most survivable mutations, have made the difference between pre-humans and humans? The answer has to be that certain of those mutations had consequences out of all proportion. Their effect was highly magnified or geared in some way, suggesting a critical effect on decisive events early in the hierarchy of cell divisions that shaped the individual.

What sort of critical effects? One at least, but more than likely a fortunate confluence of several, led to a vast enlargement of the cerebral or neopallial cortex, the region of the human brain that is so much larger proportionately than in other species. A larger brain requires a larger head, but mutations strike randomly and a mutation enhancing brain development does not grant the necessary bone growth to encase it. The folded structure of the cerebral cortex offers a clue that skull capacity may not have caught up completely with brain size. A complex and fortuitous combination of independent mutations must be envisaged, with brain enlargement and cranium expansion complementing each other, not to mention numerous other structural alterations to the skull favouring the development of speech communication. A specific gene essential for speaking has recently been isolated.

Developing the capacity for increased mental activity is not the same as exercising that capacity. A clear distinction must be drawn between instinctive behaviour patterns with which an individual is born, and memorised information or behaviour that the individual learns throughout life. The critical difference is that memorised information can never be inherited because there is no way that it can be imprinted in the genes. Numerous experiments had already proved the impossibility of inheriting any characteristic or information acquired during life, long before an understanding of DNA showed why this is so. Cutting off lambs' tails for numerous generations can have no effect on lamb DNA or newborn lambs' tails. For a few fleeting moments in the 1960s it was thought that memorised information might be encoded in RNA, following an experiment in which flatworms called planaria were taught to perform riveting tricks, like turning in response to a flash of light and an electric shock. The clever ones who learned correctly were rewarded by being chopped up and having their RNA fed to their friends who, it was claimed, learned the same tricks rather faster. The study soon failed the standard scientific test of being demonstrably repeatable by others.

A repertoire of survival behaviour is innate in every individual of a species, such as the blinking of the eyelid in response to impending eye damage, or the walking reflex that has some animals on their feet and making headway within minutes of birth. One has only to observe the behaviour of familiar animal species to be halfway misled into thinking that the young are born with a repository of knowledge, as distinct from behaviour. Compare the behaviour of cats and dogs. Cats intuitively avoid water and swimming, which many dogs enjoy. Cats are naturally inclined to run up trees, dogs are not. Dogs tend towards pack behaviour, cats are individual. The list of seemingly natural behavioural traits is endless. The quick response from those who doubt the existence of any kind of instinct is to say that the young learn these things from parents and peers. Undoubtedly many parent animals do encourage their young to catch prey, fly or whatever, but many of these habits also appear in young animals separated from their parents at birth or soon after and reared in captivity. In any case, a bird seeming to teach its fledgling to fly is demonstrating the necessity to fly and guarding its young, not explaining the highly intricate pattern of wing movements that is necessary. There is no realistic prospect that a flying insect needs instruction on how to use its priceless gift.

The riddle of apparently instinctive behaviour is far from solved, but since it is firmly established that memorised information cannot be inherited, only alternative explanations are plausible. Many behavioural responses lie in nerve–muscle circuits that process data autonomically, even in the sleeping or unconscious body. If you catch a cold and your nose becomes blocked while you are sleeping, your mouth opens to maintain breathing. The nerve circuits that organise this are genetically determined, ordained by control genes specifying their order of assembly in the developing embryo.

Genetically determined circuits are said to be hard-wired like the hardware of a computer, in contrast to the 'soft' information stored in its random access memory (RAM). The spectacular ability in some species to get up and walk almost immediately after birth requires two levels of neural input: walking as such may well be mainly a hard-wired set of reflexes (people do it in their sleep!), but the directional judgement leading the animal to follow the mother suggests an extremely quick learning of the benefits of doing so, utilising the memory part of the brain. The failures – and their genes – soon end up as lion food. Part of the initial guidance may consist of hard-wired responses towards the sounds, smells and warmth that the young has become accustomed to before birth.

As acquired characters can never be imprinted into the genes, clearly human knowledge did not cause adaptations to the brain and skull to create space for that knowledge (using the word knowledge in a broad sense to include data processing, learned reasoning processes and know-how such as mathematics). Cramming information into the brain did not cause the human head to swell (no ignoble mutterings!). The simplest evolutionary principles dictate that it had to be the other way round. The brain grew in those regions housing consciousness and memory, making possible a certain kind of mental faculty that we fondly refer to as intelligence. It is easy to mistake the accumulated knowledge of society on some topic for the accumulated knowledge stored in many brains, forgetting that it has to be transfused into each brain.

The purpose of this excursion into recent evolution is to look for clues as to how human life will continue to develop. The simple rules of molecular evolution lead to a frightening prognosis because our species, more than any other, has the power to make conscious decisions. People often talk about what 'life must have been like' many thousands of years ago. The word 'life' has two different meanings in this phraseology. It means the expression of the genome, the product being the human body with its hard-wired behavioural responses. It also means the collective society, the physical conditions and the infrastructure into which the individual was born and lived. The critical distinction is that the former, the expression of the human genome, is essentially the same now as it was many millennia ago. What may appear very superficially to be an improvement in human life over that timescale is nothing of the sort. Society as a whole may have changed its rules and its norms of behaviour, but individual humans have not changed. Society is abstract, has no palpable existence and cannot change the biological expression of life because rules of behaviour can never be transferred to the genes. The hard-wired bases of behaviour that drove Stone Age people are probably almost exactly the same today, because for one thing there has been insufficient time for significant change to have occurred, and for another any change will have been random, neutral and certainly not in response to any human desire for society to improve.

This argument works in both directions. The abilities that people now have in common were present in their early ancestors. There is little reason to doubt that virtually all abilities are present in all people, even if few have the opportunity to develop many of them. The common ancestor of all modern humans may date back at least 100 000 years. Someone born a trivial 10 000 years ago, perhaps rather more, given the appropriate education, could have trained as a jet pilot. Yet as recently as a couple of centuries ago, travel in excess of the speed of a galloping horse was deemed to be unsurvivable. In those days anyone seriously predicting the transmission of colour motion pictures to your home, let alone the ability to pull a phone out of your pocket in a Stockholm street and chat to a friend on a boat in Sydney harbour, would have been certified. All that was missing was the knowledge, the repository of information. The human mind has for tens of thousands of years possessed a tragically under-utilised potential owing to a shortage of knowledge and learned reasoning. As mentioned earlier, the brain has not developed in response to knowledge, there being no evolutionary mechanism for such a thing to happen, but knowledge has expanded in response to available brain power. Reflecting the past into the future on a similar scale, it is impossible to imagine what the human mind will be coping with in 100 000, 10 000, 1000 or even 100 years' time; if it could be imagined, it would be here now.

An axiom of the human's mental spare capacity is that nothing is to be gained by trying to select for increased intelligence. It's here already. Humans, collectively, have no difficulty in understanding new information as it is discovered. There is no reason to suspect that we are about to hit a wall of mental saturation. Indeed, a bizarre physiological condition in rare individuals, where the centre of the brain is a large void, suggests that a small fraction of the brain is sufficient for all normal purposes. Nothing seems to be standing in the way of the human brain continuing to comprehend whatever knowledge and methods of reasoning become available at the boundaries of society's encyclopaedia, given the inclination to do so. Whether individuals choose to concern themselves with such matters depends on personal choice and conditioning, which has little to do with individual mental capacity or so-called intelligence. Apart from the massive overcapacity of the brain, the neutral character of evolutionary change and its extreme slowness negate any notion of enhancing human brain power within a conceivable timescale.

If the bright side is that the human brain already has ample spare capacity, the other side is grave. The human species has the same hard-wired response to perceived threats and needs as it did 100 000 years ago. Where the response at that time may have been to smash skulls or to hunt fellow humans, so it is today. The instinctive response is hard-wired, whereas the moderating rules of society have to be impressed at every fresh generation on every single individual, who then exercises free will about whether to adopt them. A million years of teaching will not impress a single moral on a single gene and a trivial million years of future evolution will not necessarily erase a single hard-wired response to threat. Training an entire generation to respect each other's person and space will not have the slightest effect on the innate conduct of later generations. Worse, the understanding of these realities may lead to an increasing tendency to explain away conduct that was once regarded as antisocial as now genetically programmed: 'my genes made me do it'. The pre-Darwinian view of humans as elevated far above all other life took a hard knock with the discovery that we do not, after all, have a vastly superior quota of genes.

What has indeed changed at an accelerating pace over recent centuries has been the technology available for humans to implement both their instinctive and their societal responses. The combination of unerasable base instincts with the continued acceleration in technology to be expected in future centuries does not bode well for the prospects of a peaceful world. The human species with Stone Age urges on its hard disk has nuclear weapons technology loaded into its memory. In Einstein's words, 'The unleashed power of the atom has changed everything, except our mode of thinking.'

For as long as the pre-eminent species remains the human, changes can be foreseen only at the level of learned behaviour, not to the species itself. Such changes are therefore temporary. Society has no independent existence and could disappear overnight if its infrastructure collapsed. As long as it does survive, the indulgences that humanity will support are reasonably predictable: anything that increases creature comforts and self-gratification, the speed and range of communication, the speed and convenience of transportation, territorial gain or material acquisition, to name but a few. With each of these comes a heavy dependence on other people, which in turn requires ever-increasing specialisation of individuals to supply niche services. Communication underpins society as it has done since the dawn of civilisation. In the present era communication is rapidly displacing many former physical activities.

The communications revolution has already achieved a state of near-instant contact between everyone on the planet, potentially. The call centre that answers your enquiry as though across town may be located half a world away where costs are lower. More importantly, the communications revolution places the entire repository of human knowledge and information at everyone's fingertips – again, potentially. Information (or disinformation) on practically anything can be accessed from the Internet or placed there, yet the opportunities for expansion of this global resource are almost unlimited. Already encyclopaedias are hopelessly obsolete and the easiest way to find practically any information, right up to the minute, is simply to type the key words into a search engine. The unchanging but amply endowed human brain now has a rapidly expanding database with which to exercise itself. Indulgent extraterrestrials taking a peek at humanity might see us as a single gigantic brain delocalised among the entire population. If that came about, society would be critically vulnerable. Destruction of the communications infrastructure, or disinclination to spend the money to maintain it, could cause the collapse of civilization.

To be fair to space opera, the humanoid trap in alternative life-forms is indeed difficult to escape. The most detached view of life inevitably loops back to the conclusion that data processing equates with supremacy. Intelligence, in its blandest definition, combined with communication, will increasingly shape the planet. The question is whether that intelligence will be housed in humans or some unimaginable new branch of life.

If it were to be the human, since massive overcapacity is already available on the soft memory side, it is behaviour that would have to change, meaning a re-engineering of the hard-wired or read-only side. But that's not going to happen to the human, for the well-rehearsed reason that desire for it to happen cannot become imprinted on the genes. In any case, there is little sign of the desire for human behaviour to reverse out of its self-destructive course.

If the human line is doomed, as seems likely, a supreme life-form of the distant future may sprout from a much deeper and possibly new branch of the evolutionary tree. Could this happen? Of course it's not possible for evolutionary history to rewind itself and institute a new main branch of the tree. But what can happen is for an existing species to undergo mutations to key switching operations in the development of the early embryo, creating a new line with a fundamentally different body plan. This is how the lineage leading to primates and ultimately humans appeared, laying the foundations for the type of brain, bipedal habit and numerous other characteristics that easily distinguish humans from other mammals with which they share a common ancestor. It is how all the divisions of life branched away from each other.

Humans, once they are extinct, will have no say in what the next dominant division of life will be like. However worthy the vision of a perfect and harmonious world that some people possess, they have no mechanism to divert the species towards it. Evolution is not target driven and no mechanism of evolutionary nepotism can manoeuvre a close relative – a human variant or another mammal – into the vacancy. Nothing can direct evolution towards any preconceived goal, so the distant successor to humans need not be mammalian or even vertebrate, and probably won't. The desire for a revolutionary new wiring plan cannot dictate the shape of the next dominant division of life, but, looked at the other way around, a newly emerged species with a less destructive and less antisocial wiring plan could have supreme evolutionary advantages.

Who is to say that a less destructive and more sociable replacement for humans would in any sense be better? Nobody, admittedly, as evolution is the record of what happened rather than what some humans might imagine ought to happen. One could equally argue that the most destructive form of life would be the most successful (from its own point of view) because it would have the capacity to survive at all costs, destroying everything else in order to do so. Perhaps that's what will happen to humans after all. But there is an inherent reason why a less destructive offshoot of the evolutionary tree should be longer lasting, which is that by being in harmony with its environment and fellow creatures, neither need ever be destroyed, whereas humans seem to be hard-wired to destroy both.

Taken to its logical conclusion, a phenomenal increase (many orders of magnitude) in the number of hard-wired responses could begin to imitate knowledge, despite not being memorised, in the same way that a newborn giraffe seems to have the 'knowledge' of how to get up and walk, but doesn't really know anything. The key point is that hard-wired responses, comprehensive enough to mimic knowledge, are genetically transmitted and therefore uniform throughout the species. They are the nearest thing to inherited knowledge, or instinct. Noam Chomsky's theory is that language itself has a universal logic, or grammar, that is in effect part of the hard-wiring of the human brain. Recent research has found the very simplest elements of this in monkeys. Contemplate the gigantic advance from the primordial nervous reflex of protozoa to the modern human brain, which has taken most of evolutionary time to happen. Now imagine a future advance of similar magnitude. Nothing stands in the way of another truly massive increase in hard-wired responses that could equate with the emergence of a new division of life.

Humans *were* born yesterday. The entire history of hominid existence since divergence from a common ancestor with the chimpanzee, including the near-human genus *Ardipithecus*, has spanned the top one-thousandth of the evolutionary timescale on Earth. The genus *Homo*, true human, has had a still shorter existence since *Homo sapiens* appeared maybe half a million years ago – around one-ten-thousandth of life's existence. So-called civilisation extends no further back than a few thousand years or a millionth of life's existence. Medical knowledge has occupied about a ten-millionth of life's timescale and our ability to manipulate genes other than by selective breeding has appeared in the final hundred-millionth.

The present time has a dimension so fleeting that it scarcely exists. The future can never be seen and does not exist until it comes into being, which is not guaranteed. What you think you see in the present is already in the past, having happened at least an infinitesimal moment ago, or rather longer if you are gazing off the planet. What little we know about life is confined to the so-called present and the brief past of human memory and record. Life's past and future are each the answer to the other. The total impossibility of ever having extrapolated from bacteria to buffalo, or from sea-urchin to dinosaur, without hindsight, mirrors the imponderability of life's remaining billions of years on this planet. But what we cannot imagine, whether looking back or forward, has absolutely no bearing on the reality of what did happen or will happen. There never have been any constraints of time or space on life's advance and never will be. Our puny imaginations cannot shrink the problem of life's origin and development to fit on our celestially insignificant little Earth. Life belongs to the universe.

GLOSSARY

Definitions are limited to the context of this book.

ADP   adenosine diphosphate; a nucleotide containing adenine, ribose and two phosphate groups.

aerobic   with air, but often loosely used to mean with oxygen.

amino acid   formula $NH_2CH(R)COOH$ where R is a side group; 20 different examples (where R is different) are the building blocks of proteins.

anaerobic   opposite of aerobic.

anoxic   no oxygen.

argon   a non-reactive gaseous element of which a trace is present in the atmosphere.

arthropod   major division of life having jointed external skeleton, including insects, spiders, crustaceans.

astronomical unit (AU)   mean distance from the Earth to the Sun, which varies a little, standardised to 149 597 870.691 kilometres.

ATP   adenosine triphosphate; a nucleotide containing adenine, ribose and three phosphate groups.

ATPase   enzyme reacting with ATP as an energy source to catalyse another reaction.

ATP synthase (sometimes synthetase)   enzyme catalysing synthesis of ATP from ADP and phosphate.

base   component part of a nucleotide, often involved in coding genetic information (adenine, cytosine, guanine or thymine in DNA; adenine, cytosine, guanine or uracil in RNA; rarely, other bases are present).

bipedal   walking on two feet.

carbohydrate   sugars and related compounds containing carbon, hydrogen and oxygen.

carbonyl   type of compound containing carbon, oxygen and a metal.

chlorophyll   molecular combination of protein and a magnesium-containing non-protein ring structure, used by plants to trap solar energy.

chloroplast   chlorophyll-containing organelle in plant cells (many per cell).

chromosome   microscopically visible structure in the nucleus of a cell, comprising the DNA for many genes; the human cell has 46 chromosomes (23 in sperm or egg).

codon the length of RNA (3 nucleotides) coding for one amino acid in the sequence of a protein (or an instruction to begin or terminate the protein).

collagen a fibrous protein giving strength to skin and connective tissue.

cyanobacteria blue-green bacteria with photosynthetic capability.

cytochrome molecular combination of protein and a non-protein ring structure containing metal (usually iron), acting to convey electrons for oxidation and reduction.

dehydrogenase enzyme involved in catalysing oxidation and reduction by transferring hydrogen.

DNA deoxyribonucleic acid, a polymer of nucleotides; the sugar in the component nucleotides is deoxyribose.

entropy term in thermodynamics for disorder or the unavailability of energy.

enzyme catalytic protein (name usually ends in -*ase*).

eukaryote cell type more complex than prokaryote (bacteria) with membrane-bound nucleus, internal organelles and other distinctions.

exon length of DNA expressed ultimately as protein (or special RNA); *see* intron.

expression process in which the information content of a gene is directed into making a protein (or a special RNA).

formaldehyde simple organic molecule, formula HCHO.

gene the unit of DNA expressed as a single protein (or special RNA).

genome the entire complement of genes for a species (strictly, an individual of the species).

globin the protein component of haemoglobin or myoglobin.

glycine the simplest amino acid, NH_2CH_2COOH.

GTP guanosine triphosphate; a nucleotide containing guanine, ribose and three phosphate groups; *compare with* ATP.

haem often spelt heme, a particular ring structure containing iron that associates with globin to form functional haemoglobin.

haemoglobin often spelt hemoglobin, a globin–haem molecule that binds oxygen reversibly.

hafnium rare metallic element, atomic number 72.

haplosome simple predecessor of the ribosome.

hard-wired nerve circuit enabling a particular response, such as the knee-jerk.

Heisenberg's uncertainty principle the more precisely the position of a particle is determined, the less precisely its momentum is known and *vice versa*; in other words, one cannot measure both with absolute precision, or know everything about a particle.

helix many biological molecules including parts of proteins, collagen (a protein) and DNA are helical in shape; the term double helix refers *only* to DNA having two strands twisted around each other.

hexokinase enzyme catalysing the conversion of glucose and ATP to glucose phosphate and ADP.

histone bobbin-like protein around which DNA is wound in the nucleus.

Hubble constant H_0 currently measured relationship between the distance of a star from the observer and its velocity (commonly expressed as km/sec per megaparsec of distance, which simplifies to inverse time).

hydrocarbon molecular chain of carbon atoms, each with hydrogens attached.

hydrophilic chemical character favouring contact with water rather than oil or fat.

hydrophobic chemical character favouring contact with oil or fat rather than water.

hypoxic low oxygen.

impermeable of a membrane, not allowing the molecule in question to pass through.

intron intervening sequence, or length of DNA not ultimately expressed as protein or special RNA; *see* exon.

invertebrate not having the skeleton and characteristics of a vertebrate, especially a vertebral column; *see* vertebrate.

isotope variant of a chemical element having essentially the same chemical character and atomic number but a slightly different atomic weight.

Kelvin (K) temperature scale starting from the (unattainable) absolute zero, having divisions the same size as degrees Celsius; zero K (the ° symbol is not used) is close to -273.16 °C, so 0 °C is 273.16 K.

keratin protein found in hair, nails, feathers and other structures.

kinetic energy energy of motion.

light, speed in a vacuum (symbol c) = 299 792 458 metres per second (by definition; other units are derived from it).

light-year distance travelled by light in a vacuum in 1 standard year (365.2425 days). A light-year is by definition 9 460 536 207 068.016 kilometres, or 3.0659458×10^{-7} megaparsecs; *see* parallax and megaparsec.

lipid fatty molecule.

lipoprotein protein with lipid molecules attached.

lutetium rare metallic element, atomic number 71.

lysosome organelle of eukaryotic cells containing numerous enzymes that catalyse breakdown of particular biochemicals.

lysozyme an enzyme found e.g. in eye tears, having a protective role by attacking bacteria.

megaparsec (Mpc) a million parsecs; *see* parallax and light-year.

metamorphosis change of form, e.g. from tadpole to frog.

mitochondrion (plural mitochondria) organelle in eukaryotic cells having numerous functions including the generation of ATP from ADP and phosphate, making use of energy from the oxidation of hydrogen to water (many mitochondria per cell).

mutant (adjective or noun) changed by a mutation; strictly happens in DNA, but colloquially can refer to DNA, protein, individual or species.

mutation change to a gene DNA, the effect of which can be unfelt if the expressed protein is unchanged, or may cause a substituted amino acid, or more far-reaching consequences such as the disruption or inactivation of a gene.

myoglobin fairly similar to haemoglobin but having primarily an oxygen-storage rather than an oxygen-transport role.

NADP nicotinamide adenine dinucleotide phosphate, a hydrogen-transferring molecule that works in conjunction with a dehydrogenase enzyme.

nanotechnology extremely small device (from nano- meaning 10^{-9}, being smaller than micro- meaning 10^{-6}).

neodymium rare metallic element, atomic number 60.

nucleotide usually a component of a nucleic acid (RNA or DNA) containing a base, sugar (ribose or deoxyribose) and phosphate. Some nucleotides such as ATP, GTP and NADP function independently of nucleic acids.

nucleus (of cell) membrane-bound organelle containing the chromosomes.

organelle component part of a cell such as mitochondrion, chloroplast, lysosome, nucleus.

osmosis physical phenomenon resulting in water transport across a membrane through a tendency for concentrations to equilibrate on both sides.

oxidase enzyme catalysing oxidation.

oxido-reductase enzyme catalysing mutual oxidation of one molecule and reduction of another.

ozone molecule of three oxygen atoms rather than the usual two.

panspermia early concept of the elements of life being scattered throughout space.

parallax, parallax-second, parsec parallax is the apparent angular displacement of a target. A star that appeared to move by one angular second (1/3600 of a degree) between sightings on a baseline of one astronomical unit (see above) is, by geometry, at a distance of 30 900 billion kilometres or 3.26 light years. On this basis the parsec is a unit of distance; *see* light-year and megaparsec.

permeable of a membrane, allowing the molecule in question to pass through.

plutonium metallic radioactive element atomic number 94, not now occurring naturally on Earth but made in nuclear reactors.

polymerisation joining molecules together, as the joining of amino acids to form protein or of nucleotides to form DNA or RNA.

polysaccharide polymer of sugars, such as starch, which is made of chains of glucose.

porphyrin a large non-protein molecule containing four rings and a site that holds a metal atom, as present in haemoglobins, cytochromes and chlorophylls.

prokaryote bacterial type of cell, with no nuclear membrane around the genes and having other differences from eukaryotes.

protein polymer comprising a sequence of amino acids.

protozoon (plural protozoa) single-celled eukaryotic organism.

radio-isotope an isotope that is radioactive; many elements may have some radioactive and some non-radioactive (stable) isotopes.

radium rare metallic element, atomic number 88.

red shift shift of colour of light towards a lower frequency or longer wavelength (i.e. 'redder') owing to the source moving away from the observer, analogous to the change in pitch as a train whistle or siren passes.

ribosome small particulate body in cell (thousands present) where mRNA is translated and proteins are assembled.

RNA ribonucleic acid, a polymer; the sugar in the component nucleotides is ribose.

   mRNA messenger RNA, carries coded information transferred from DNA for eventual translation into protein.

   rRNA ribosomal RNA, not translated into protein but forming part of the structure of the ribosome.

   tRNA a short polymer about 80 nucleotides in length, not translated into protein but functioning to carry amino acids to their reaction on the ribosome.

   'special RNA' in this glossary means rRNA or tRNA (i.e. non-translated RNA).

rubidium rare metallic element, atomic number 37.

rubisco ribulose bis-phosphate carboxylase-oxygenase, an enzyme catalysing the incorporation of carbon dioxide into plants.

samarium rare metallic element, atomic number 62.

spectrometer, spectroscope, spectrograph instruments for measuring or viewing the colours (wavelengths or frequencies) of light.

spectrum the frequencies (wavelengths, colours) corresponding to light, or a display of these from a particular source such as the Sun or a star.

strontium metallic element, atomic number 38.

substrate a molecule that will be converted to a product by enzyme catalysis.

supernova large new star, detected as a massive release of light and other energy.

symbiosis living together with some discernable mutual advantage.

synthase enzyme catalysing synthesis (particularly ATP synthase, see above).

thermodynamics transformations of heat and energy, the important laws governing these processes, and the consequences for the movement of atoms and molecules.

thermonuclear the energy-releasing process of nuclear fusion going on in the Sun or a star.

thorium metallic element, atomic number 90.

translation conversion of the information content of mRNA (transferred from DNA) into protein.

vertebrate having an internal skeleton and other characteristics, particularly a vertebral column enclosing part of a central nervous system; *see* invertebrate.

zircon mineral (minor gem) containing the rare metallic element zirconium, atomic number 40.

zygote cell resulting from the fusion of sperm and egg cells.

UNITS AND DIMENSIONS

Angström 10^{-10} metre.

billion 10^9 (in earlier English as distinct from American usage it could mean 10^{12}).

kilocalorie (kcal) 1000 calories; multiply kcal by 4.18 to convert to kilojoules.

kilojoule (kJ) 1000 joules; divide kJ by 4.18 to obtain kilocalories. (Calories originated in studies of heat, joules in studies of energy, but these are related through the mechanical equivalent of heat, hence the simple conversion constant of 4.18.)

light-year distance travelled by light in vacuum in one year (see above).

micro (microlitre etc.) one-millionth, 10^{-6}.

μ 10^{-6} (sometimes used alone to mean μm, see below).

μm 10^{-6} metre, micrometre, also called a micron.

mμ milliμ, 10^{-9} metre (i.e. $10^{-3} \times 10^{-6}$); the more modern nm (nanometre, see below) avoids confusion between μm and mμ.

Mpc megaparsec (distance of 3.09×10^{19} km).

nano (nanometre etc.) 10^{-9}.

pico (picogram etc.) 10^{-12}.

Appendix

This Appendix contains some calculations and sources of additional information. Calculations of huge consequence are often astonishingly simple, as you will see, so give them a try. Scientific debate usually revolves not around the calculation itself, but around the correctness of the numbers put into the calculation.

References to further information are to short, readable news items where possible. The time-honoured convention is for original scientific research to be reported in journals (or sometimes the 'proceedings' of a conference) after scrutiny by an editor and independent scientists, who carefully check whether the conclusions are justified by the methods used. Were the controls adequate, are the results statistically valid, have the conclusions discussed and acknowledged previous work on the topic? This is the process known as peer review, but it is far from failure proof. Such articles are the primary sources of scientific information, but unfortunately they are often written so turgidly as to be practically unintelligible to the non-scientist, or even to the scientist unfamiliar with the jargon of the same narrow field. Conventions of format make the paper about twenty times longer than the punch line discovery being reported.

Publications such as *New Scientist* select interesting snippets and present them in an intelligible and newsy manner, but also tell the reader the names and institutions involved and where to find the primary paper. Some of the primary journals such as *Nature* and *Science* devote a few pages to general scientific news, with articles in which a good communicator primes readers about the significance of a detailed paper in the same issue. The references in this Appendix are generally to secondary sources of these kinds, but lead quickly to primary sources if the reader needs them.

CHAPTER 1

The basis of Darwin's age of 306 million years for the Weald appears to be as follows, shown first in original imperial miles, feet and inches (giving Darwin's exact result), then as the approximate metric equivalent.

1 IMPERIAL MEASURE

Erosion of a 500-foot cliff was assumed to be 1 inch per century. Erosion distance 22 miles, each mile is 5280 feet, each foot is 12 inches. Original height of the Weald cliff taken to be 1100 feet.

$$\frac{\text{Erosion}}{\text{time}} = 22 \times 5280 \times \frac{12}{1} \times \frac{1100}{500} = \begin{array}{l} 3.066624 \text{ million centuries} \\ (306\ 662\ 400 \text{ years, or} \\ \text{about 300 million years}) \end{array}$$

2 METRIC CONVERSION

Erosion of a 150-metre cliff assumed to be 25 millimetres per century. Erosion distance 35 km, and there are 10^6 (a million) millimetres in a kilometre. Original height of the Weald cliff taken to be 330 metres.

$$\frac{\text{Erosion}}{\text{time}} = \frac{35 \times 10^6}{25} \times \frac{330}{150} = \begin{array}{l} 3.08 \text{ million centuries} \\ (308 \text{ million years}) \end{array}$$

Reference to Laing (also mentioned in Chapter 9): Laing, S. (1890) *Problems of the Future and Essays*, London: Chapman and Hall.

Excellent commentaries on the historical studies of the age of the Earth are to be found in Burchfield, J.D. (1975) *Lord Kelvin and the Age of the Earth*, London: Macmillan Press; and Dalrymple, G.B. (1991) *The Age of the Earth*, Stanford, CA: Stanford University Press (rich in further references).

A *Nature* article on the temperature history (thermochronometry) of asteroids is summarised by John A. Wood in the same issue (2003) 'Of asteroids and onions', *Nature*, 422: 479–480.

The role of zircons in determining Earth's early history is summarised by Halliday, Alex N. (2001) 'In the beginning ...', *Nature*, 409: 144–145 (referencing a research paper in the same issue). Larry O'Hanlon also comments on the finding in *New Scientist*, 20 April 2002, p 18.

A good review article on the flat, closed and open models of the universe is Bahcall, N.A., Ostriker, J.P. and Steinhardt, P.J. (1999) 'The cosmic triangle: revealing the state of the universe', *Science*, 284: 1481–1488.

For a possible acceleration in the expansion of the universe, see the brief news item by Sharmila Kamat, 'Universe stepped on the gas 5 billion years ago', *New Scientist*, 18 October 2003, p 17.

For more detail, see Hogan, C.J., Kirshner, R.P. and Suntzeff, N.B. (1999) 'Surveying space-time with supernovae', *Scientific American*, January: 46–51.

CALCULATION OF THE AGE OF THE UNIVERSE FROM HUBBLE'S CONSTANT (H_0), WHICH IS THE RATE OF EXPANSION DERIVED FROM RED SHIFTS

Thanks to a stroke of luck with two numbers being so close that they cancel out (3.09 and 3.15, see below), there is a short cut to converting from H_0 to universe age in billions of years: simply divide 1000 by H_0. For example, if $H_0 = 75$, then age $= 1000/75$, which is 13.3 billion years (provided age really is the inverse of H_0 – see below).

Here is the calculation in full. Assume a consensus value for H_0 of 75 kilometres per second per megaparsec (km.s^{-1}.Mpc^{-1}). One megaparsec (Mpc) is a measure of distance equal to 3.09×10^{19} km, say 3×10^{19} km in round figures.

$$\text{Therefore } H_0 = \frac{75 \text{ km per second}}{3 \times 10^{19} \text{ km}} \qquad (\text{i.e. km.s}^{-1}.\text{km}^{-1})$$

since km on both lines cancel each other, that is the same as

$$= \frac{75}{3 \times 10^{19}} \qquad \text{per second (s}^{-1})$$

$$= \frac{25}{10^{19}} \quad \text{s}^{-1}$$

Since s^{-1} means per second, inverting the above gives a value in seconds:

$$\text{age of universe} = \frac{10^{19}}{25} \qquad \text{seconds}$$

which comes to 4×10^{17} seconds as the age of the universe.

Dividing 4×10^{17} seconds by the number of seconds in a year, which is 3.15×10^7 (say 3×10^7) gives 1.3×10^{10} years, or *13 billion years*.

The above calculation is based on the consensus figure of $H_0 = 75$. Alternative results based on different values of H_0 are:

Taking the lower accepted figure of $H_0 = 50$ gives 20 billion years.
Taking the upper accepted figure of $H_0 = 100$ gives 10 billion years.

Another way of finding the limit of the visible universe from H_0 is as follows. If $H_0 = 75$ km/s per Mpc, recession is 75 km/s at a distance of 1 Mpc. Recession would equal the speed of light

(300 000 km/s) at a distance of 300 000/75 = 4000 Mpc. Since a light year is close to 3×10^{-7} megaparsecs, $4000/(3 \times 10^{-7})$ is 13.3 billion light-years. Light that has been travelling towards us for longer has not yet arrived.

This calculation is not necessarily as cut and dried as is usually presented, because it depends on a cosmological model in which the rate of expansion has been constant and age $= 1/H_0$. According to some models of the universe the relationship between H_0 and the age of the universe may be more complex than a simple $1/H_0$, such as $2/(3H_0)$, although the result would still be in the region of billions of years.

UNITS OF THE WAVELENGTH OF LIGHT

The wavelength of light is the distance between wave peaks, typically 400–700 nanometres depending on colour. A nanometre (nm) is 10^{-9} metre, i.e. a billionth (thousand-millionth) of a metre. In older books the unit may be written mμ, meaning milli-micron, where milli means 10^{-3} and a micron μ is 10^{-6} metre; mμ is therefore the same size as nm (10^{-9} metre). An alternative unit sometimes used is the Angström (written Å), which is 10^{-10} metre or one-tenth the size of nm, so that 400 nm would be 4000 Å.

The speed of light in a vacuum can be taken as 300 000 km per second or 3×10^8 metres per second. That's near enough to the exact figure of 299 792 458 metres per second. The length of a metre is now derived from the speed of light, and not *vice versa*, so if in the future any adjustment becomes necessary, the length of the metre will change (infinitesimally) and the speed of light will keep the same value.

A recent experimental confirmation that light travelling long distances does not become tired with time is reported briefly in *Science* (2001) 292: 2414.

Dating the universe by the alternative radioactive cosmo-chronology method is the subject of a research paper in *Nature* that is well interpreted by Christopher Sneden in an explanatory article in the same issue: (2001) 'The age of the universe', *Nature*, 409: 673–675.

CHAPTER 2

Piltdown Man: Many people have been considered as possible culprits in the fascinating Piltdown hoax story and there have been several books published. Richard Harter maintains an excellent resource on the Internet, which leads to many more resources. Since private Web addresses change it is unhelpful to cite them, but keywords Piltdown and Richard Harter should find it quickly (and indeed many others).

See also Vines, G. (2003) 'Toad in the hole', *New Scientist*, 9 August: 50–51. This is about Piltdown Man and another forgery that adds strength to the idea that the forger may have been Charles Dawson.

Kinetic energy on impact is $\frac{1}{2}mv^2$ where m is mass and v is velocity, the effect of which is squared, which is why the effect of speed in a crash is very nasty. Jumping from twice the height increases impact speed (v) by a factor of $\sqrt{2}$ (about 1.4), the square of which is of course 2, doubling kinetic energy.

For ideas about birds becoming dinosaurs, see Gregory S. Paul (2002) *Dinosaurs of the Air*, Baltimore, MA: John Hopkins University Press, and (1988) *Predatory Dinosaurs of the World*, New York: Simon and Schuster.

For a summary of anatomical features associated with the development of flight in birds, see Shipman, P. (1997) 'Birds do it ... did dinosaurs?' *New Scientist*, 1 February: 27–31 and Shipman, P. (1998) *Taking Wing: Archaeopteryx and the Evolution of Bird Flight*, New York: Simon & Schuster.

News items on what may be hindleg feathers of *Microraptor*, *Archaeopteryx* and other ancient birds are in *New Scientist*,

25 October 2003, p 15 and in *New Scientist*, 25 January 2003, pp 14–15.

CHAPTER 3

Junk DNA, see Gibbs, W.W. (2003) 'Unseen genome: gems among the junk', *Scientific American*, November: 27–33.

Numerous editions of Charles Darwin's *Origin of Species* (*On the Origin of Species by Means of Natural Selection or the Preservation of Favoured Races in the Struggle for Life*) have been published, including Darwin's own revisions. My particular source of reference is an edition published in 1903 for the Rationalist Press Association by Watts & Co, Fleet Street, London, and said by the publisher to be 'an exact reprint of the first edition (published 1859–60)'. One quick check on whether you have a reprint of the first, unrevised edition (or a priceless original, should you be so lucky) is a reference to the Creator in the last half-dozen lines of the first edition: 'There is grandeur in this view of life, with its several powers, having been originally breathed by the Creator into a few forms or into one ...'

The unfavourable comment on Chambers' (then anonymous) *Vestiges of the Natural History of Creation* quoted from the *Scientific American* of 1852 was reproduced in the February 2002 issue of the *Scientific American*, p 8, as a reflection on 150 years ago.

The Leclerc (de Buffon) quotation is reproduced from *Before the Deluge* by Herbert Wendt (1968), New York: Doubleday, p 79.

The report on *Colymbosathon ecplecticos* as both the earliest soft tissue impression yet imaged and the remarkably little anatomical change in 425 million years, is briefly reported by Erik Stokstad (2003) in *Science*, 302: 1645, with reference to a full article in the same issue.

Table A.1 The Genetic Code

FIRST LETTER ⇓	SECOND LETTER U	C	A	G	THIRD LETTER ⇓
U	Phenylalanine	Serine	Tyrosine	Cysteine	U
U	Phenylalanine	Serine	Tyrosine	Cysteine	C
U	Leucine	Serine	STOP	STOP	A
U	Leucine	Serine	STOP	Tryptophan	G
C	Leucine	Proline	Histidine	Arginine	U
C	Leucine	Proline	Histidine	Arginine	C
C	Leucine	Proline	Glutamine	Arginine	A
C	Leucine	Proline	Glutamine	Arginine	G
A	Isoleucine	Threonine	Asparagine	Serine	U
A	Isoleucine	Threonine	Asparagine	Serine	C
A	Isoleucine	Threonine	Lysine	Arginine	A
A	Methionine or START	Threonine	Lysine	Arginine	G
G	Valine	Alanine	Aspartate	Glycine	U
G	Valine	Alanine	Aspartate	Glycine	C
G	Valine	Alanine	Glutamate	Glycine	A
G	Valine	Alanine	Glutamate	Glycine	G

Each amino acid is coded for by a sequence of three nucleotides, shown here by their initial letters: U, C, A or G. The three nucleotides make up a codon. For example, if the first is U, the second is G and the third is G, the codon UGG codes for the amino acid tryptophan. Three of the codons (UAA, UAG, UGA) do not code for an amino acid but signify STOP, meaning the end of the protein. The codon AUG codes for methionine but can also signal the START of a protein.

The abbreviations U, C, A and G stand for the nucleotides present in an RNA sequence and for the bases they contain: uracil, cytosine, adenine and guanine. The coding strand of DNA has the same sequence of bases as RNA, except that DNA has thymine (T) where RNA has uracil (DNA also contains long lengths of non-coding sequence between genes and sometimes within them).

RNA is a single sequence. DNA has two strands in which the coding strand and the template strand are wrapped around each other to form the double helix, with A of one strand always in contact with T of the other, and G of one in contact with C of the other.

The genetic code is *almost* universal. Slight differences are found in, for example, mitochondria.

The table of the genetic code traditionally shows the RNA version and therefore contains U, but sometimes it is written with a T and by implication refers to DNA. Pretending for a moment that T and U are the same (they dictate the same information), the RNA sequence is the same as one of the DNA strands (called the coding strand) and the opposite of the other (the template strand). For the beginning of the human insulin molecule the sequences are:

AAA CAC TTG GTT ...   (DNA template strand sequence)
TTT GTG AAC CAA ...   (DNA coding strand sequence, the 'opposite' of above)
UUU GUG AAC CAA ...   (messenger RNA sequence, 'same' as above, but U replaces T)
Phe Val Asn Gln ...   (amino acid sequence of protein, translation of above)

CHAPTER 4

The formula for the Poisson distribution is:

$$\frac{\boldsymbol{m}}{100} = -\ln\left(\frac{1 - \boldsymbol{n}}{100}\right)$$

where $\boldsymbol{n}$ is the observed number of differences per 100 amino acids and $\boldsymbol{m}$ is the true historical number; ln means logarithm to base $\boldsymbol{e}$.

The graph (Figure 4.1), based on the work of Margaret Dayhoff, is redrawn from Dickerson, R.E. and Geiss, I. (1983) *Hemoglobin: Structure, Function, Evolution and Pathology*, San Franciso, CA: Benjamin Cummings, p 92.

A recently discovered artefact that could account for apparent fossil bacteria is mentioned in a brief article: Hogan, J. (2003) '"Microfossils" made in the laboratory', *New Scientist*, 22 November: 14–15.

For further information on the ^{13}C signature of early life it is necessary to pursue some of the full journal articles, which themselves contain many other references. For example, Rosing, M.T. (1999) '[13]C-depleted carbon microparticles in >3700-Ma seafloor sedimentary rocks from West Greenland', *Science*, 283: 674–676; Schidlowski, M. (1988) 'A 3800-million-year isotopic record of life from carbon in sedimentary rocks', *Nature*, 333: 313–318; Mojzsis, S.J., Arrhenius, G., McKeegan, K.D., Harrison, T.M., Nutman, A.P. and Friend, C.R.L. (1996) 'Evidence for life on Earth before 3800 million years ago', *Nature*, 384: 55–59.

A research paper on the [13]C (and ^{18}O) composition of squid 'bone' (i.e. shell), including some methodology, is to be found in Bettencourt, V. and Guerra, A. (1999) 'Carbon- and oxygen-isotope composition of the cuttlebone of *Sepia officinalis*: A tool for predicting ecological information?' *Marine Biology*, 133: 651–657.

Not everyone accepts that lifelike signs in ancient rocks have anything to do with life; for contrary views see Fedo, C.M. and

Whitehouse, M.J. (2002) 'Metasomatic origin of quartz-pyroxene rock, Akilia, Greenland, and implications for Earth's earliest life', *Science*, 296: 1448–1452; Bridgwater, D., Allaart, J.H., Schopf, J.W., Klein, C., Walter, M.R., Barghoorn, E.S., Strother, P., Knoll, A.H. and Gorman, B.E. (1981) 'Microfossil-like objects from the Archaean of Greenland: A cautionary note', *Nature*, 289: 51–53; Roedder, E. (1981) 'Are the 3800-Myr-old Isua objects microfossils, limonite-stained fluid inclusions, or neither?' *Nature*, 293: 459–462.

A key paper providing evidence of early photosynthesis from ancient preserved organic chemicals is Brocks, J.J., Logan, G.A., Buick, R. and Summons, R.E. (1999) 'Archean molecular fossils and the early rise of eukaryotes', *Science*, 285: 1033–1036; a short commentary by Andrew H. Knoll is in the same issue: *Science*, 285: 1025–1026.

CHAPTER 5

For a recent summary of the debate over the date of appearance of oxygen in the atmosphere, see the news article by Jon Copley (2001) in *Nature*, 410: 862–864.

See also the news item with references by Stephen J. Mojzsis (2003) 'Probing early atmospheres', *Nature*, 425: 249–251.

The mechanism by which the rotating protein motor generates ATP is explored in detail in the review article by Ostler, G. and Wang, H. (2003) 'Rotary protein motors', *Trends in Cell Biology*, 13: 114–121.

CHAPTER 6

Formation of complex organic molecules in space: a recent paper is summarised by Jenny Hogan in 'Mystery particle floats in from deep space', *New Scientist*, 6 March 2004, p. 9.

For a news item by Peter Aldhous on the extreme slowness of spontaneous reactions such as would lead to protein formation, see *New Scientist*, 25 February 1995, p 16.

The ability of mineral surfaces to act as catalysts, particularly in the building of nucleic acid precursors, has been well studied by James Ferris and Leslie Orgel and is widely reported. The contribution of radioactive elements is referenced in a news item by Graham Lawton (2003) 'Oily grains kick-started life on Earth', *New Scientist*, 12 April, p 23.

Work on the synthesis of chemicals in simulated space conditions is reported by Gretel Schueller (1998) in *New Scientist*, 12 September, pp 30–35, and by Everett L. Schock (2002) in *Nature*, pp 380–381. For more detail see in the same issue of *Nature*: Bernstein, M.P., Dworkin, J.P., Sandford, S.A., Cooper, G.W. and Allamandola, L.J. (2002) 'Racemic amino acids from the ultraviolet photolysis of interstellar ice analogues', *Nature*, 416: 401–403. Also Muñoz Caro, G.M., Meierhenrich, U.J., Schutte, W.A., Barbier, B., Arcones Segovia, A., Rosenbauer, H., Thiemann, W.H.-P., Brack, A. and Greenberg, J.M. (2002) 'Amino acids from ultraviolet irradiation of interstellar ice analogues', *Nature*, 416: 403–406.

Deep-sea conversion of nitrogen to ammonia is described in Brandes, J.A., Boctor, N.Z., Cody, G.D., Cooper, B.A., Hazen, R.M. and Yoder, S.Y. (1998) 'Abiotic nitrogen reduction on the early Earth', *Nature*, 395: 365–367.

Figures for the amounts of material deposited in the atmosphere from the vaporisation of asteroids and comet debris are obviously difficult to be exact about. The figure of about 40 000 tonnes per year commonly quoted – e.g. Don Brownlee, *Nature*, 395: 113 (1998) – has been around for some time, a similar amount having been cited as long ago as 1964 by Boschke, F.L. in *Creation Still Goes On*, London: Hodder and Stoughton.

A detailed review of the similarities between vertebrate and invertebrate hormone molecules by De Loof, A. and Schoofs, L., is in the research journal *Comparative Biochemistry and Physiology* (1990) 95B: 459–468.

CHAPTER 8

The textbook volume of Earth's oceans is 1347 million cubic kilometres, i.e. approximately 1.3×10^9 km^3. Each km^3 is 1 000 000 000 000 litres (10^{12} L) or 1 000 000 000 000 000 millilitres (10^{15} mL), therefore the volume of the oceans is $1.3 \times 10^9 \times 10^{15}$ mL, which is 1.3×10^{24} mL, say about 10^{24} mL.

MATTER REACHING THE EARTH FROM SPACE

The difference between a meteor and a meteorite is that a meteor or 'shooting star' is what is seen when a piece of matter is heated to glowing during passage through the atmosphere; a meteorite is the physical object that hits the Earth.

The amount of material in the Oort cloud of cometary material is argued in Stern, S.A. and Weiaaman, P.R. (2001) 'Rapid collisional evolution of comets during the formation of the Oort cloud', *Nature*, 409: 589–591.

See also Bailey, M.E. (2002) 'Where have all the comets gone?' *Science*, 296: 2151–2153.

Recent research on the origin of the Kuiper belt is discussed in Gomes, R. (2003) 'Conveyed to the Kuiper belt', *Nature*, 426: 393–394.

Dust, asteroid and comet material hitting the Earth: Taylor, A.D., Baggaley, W.J. and Steel, D.I. (1996) 'Discovery of interstellar dust entering the Earth's atmosphere', *Nature*, 380: 323–325; Hughes, D.W. (1996) 'Dust from beyond the solar system', *Nature*, 380: 283; Frank, L.A. (1990) *The Big Splash*, New York: Carol Publishing Group (ice comets).

Nicolas Dauphas's work on comets and Earth's water and atmosphere is referenced in a news item in *New Scientist*, 11 October 2003, p 17.

Brownlee, D. (1998) 'Ancient cosmic spherules', *Nature*, 395: 113–115.

Pictures of the spectacular impact of comet Shoemaker–Levy 9 in 1994 can be seen on the Internet site http://nssdc.gsfc.nasa.gov/planetary/comet.html if still available; otherwise run a search for Shoemaker–Levy.

Close-up photographs of asteroids can be found on NASA Web sites. A series of articles on the NEAR-Shoemaker spacecraft encounter with Eros in 2001 appeared in *Nature*, 27 September 2001 (413: 369–370, summary, and 390–400).

For a short but excellent biography of the late Eugene Shoemaker, see http://wwwflag.wr.usgs.gov/USGSFlag/Space/Shoemaker/

Cometary material in asteroids: see brief note with more detailed reference; 'Solar system's damp start written in asteroid rock', *New Scientist*, 6 March 2004, p 17.

Interstellar transport of microbes, brief note by Hecht, J. (2001) 'Life will find a way', *New Scientist*, 17 March, p 4.

Asish Basu's work on the Permian extinction is summarised in a news item in *New Scientist*, 20 November 2003, p 15, with reference to the definitive article in *Science*, 302: 1388–1392.

For a reference to the Lake Acraman asteroid impact see the review by Rachel Nowak (2003) 'Rock and ice vie for credit for complex life', *New Scientist*, 3 May, p 17.

The gravitational attraction (F) between two bodies depends on their masses (large body M, small body m), the distance between their centres of gravity (d), and the gravitational constant (G):

$$F = G \frac{Mm}{d^2}$$

The value of G is 6.67300×10^{-11} m^3 kg^{-1} s^{-2} (or N m^2 kg^{-2}).

Martian atmosphere and when it may have been lost: see the brief article by Eugenie Samuel in *New Scientist*, 10 February 2001, p 4.

CHANCES OF COMMUNICATION WITH ALIENS ON ANOTHER PLANET

The various calculations are always so arbitrary as to be practically futile. Very briefly and for the sake of something to dispute, try this.

A radio signal broadcast in all directions decreases in strength with the square of the distance travelled (the inverse square law), soon becoming vanishingly weak. On the scale of interstellar distance, the power required for omnidirectional communication is far beyond contemplation. As with starlight itself, it often requires the entire energy output of a star – often millions of times more powerful than our own Sun – for a faint signal to reach us. The only option is to transmit all the available power in one direction, but there is no point in aiming the signal at any particular solar system because it is impossible to know where it will be at the time of arrival. The same problem has confronted any civilisation wishing to contact us. It could not know our future position, so there is no possibility of a signal being intentionally directed at us.

There are some subtleties in this: certainly we can point an antenna at what seems to be the position of a specific star system and be sure it is the direction *from* which its light, or any radio signal, has come. However, the star is not there any more, having moved since the light was transmitted. There is also the possibility that the light did not really come from the star it seems to have, but came from another star closer or more distant along the same line of sight. Therefore it is useless to reply along the same line of sight and we have insufficient information to predict the star's future position at the time when the reply would arrive. For the same reasons a signal cannot have been sent *to* us intentionally because we are moving, unless the distant civilisation had some superior knowledge of our future position.

The solution is to send a narrow beam in as many different directions as possible in the hope of its being intercepted by someone. This is a notional solution because it wouldn't really work. The directional beam must be so narrow as to be effectively parallel, which is also pretty much impossible. Visualise it this way. An angle of one degree sounds really narrow; yet the apparent width of the Moon, as seen from the Earth, represents only half a degree. Radio waves take only one and a quarter seconds to travel from the Earth to the Moon. If a radio beam from Earth a mere half-degree wide has spread out to the diameter of the Moon after one and a quarter seconds, then how wide, how incredibly feeble, would the beam be after plodding on for a million years? Or a billion years?

A half-degree beam is clearly no use for the purpose, but the best microwave technology can transmit a beam so nearly parallel that it spreads about 3 cm at a distance of 10 000 km. That's less than a millionth of a degree, so a beam about a millionth of a degree wide might be a fair model of what could be done. Such a beam would have widened to about 10 times the width of an Earth-like planet after 1000 light-years, to about 10 000 times the width after a million light-years and to about 10 000 000 000 times the width of a planet after a billion light-years. At that point the distance travelled would still be less than a tenth of the distance to the remotest visible stars.

The same beam would need to be pointed in about 10^{17} different directions in turn to cover the whole sky. This approximation is obtained by envisaging a transmitting antenna that transmits a near-parallel beam for a short burst (say one second) through every successive millionth of a degree for 360° in a single plane, or disk, like a thin slice out of the centre of an orange. The disk is then turned on edge by a millionth of a degree and the process repeated until it has again swept a full 360°. Every slice would transmit

360×10^6 signal bursts, and the number of slices would be 360×10^6. The total number of directions possible is the product of these two numbers, about 10^{17}. Rationing each burst to just one second would occupy 3 billion years – nearly as long as the Earth has existed and vastly longer than civilisation has existed.

The humanoids at the other end have exactly the same problem in pointing their narrow-beam receiving antenna. Once every 3 billion years they listen in our direction for just one second. And once in 3 billion years we are transmitting in their direction – but not at the same moment – we missed each other! The arithmetic tells us that at this rate, one-way communication with both antennas pointing at each other will be established for one second every 10^{19} years (3 billion $\times$ 3 billion).

Maybe this is unduly pessimistic. Confining the search to the densest part of the Milky Way greatly reduces the survey area and improves the chances by maybe a factor of 10^5. Perhaps our interstellar correspondents have such a powerful transmitter that they can use a beam a hundred times wider, which multiplies the chance of intercepting it by another 10^4. A cunning new receiver waiting to be invented is so selective and sensitive that it helps by another 10^2. Where does all this optimism get us? An improvement factor of 10^{11}; now we have to wait only 100 000 000 years for our one second of hot gossip. And this assumes a degree of clairvoyance that inspired both civilisations to use precisely the same wavelength!

All this is just a matter of kicking figures around and nobody knows any answers. If the beam is sharpened towards being perfectly parallel, then it takes longer to sweep the sky. Widening the beam to shorten the sweep time makes the signal strength vastly weaker, whether sending or receiving. You can't win.

If, nevertheless, you would like to donate some spare time to the search for extraterrestrial intelligence and perhaps become the most famous person in the history of life on Earth, read for initial instructions: 'If ET's out there, we'll find him', *New Scientist*, 20 July 2002, p 22.

Work on an obscure property of light that could enable information to be encoded in a novel way is summarised by George Musser. There is even an experiment you can download from the Internet. See Musser, G. (2003) 'All screwed up. An obscure property of light puts a spin on astronomy', *Scientific American*, November: 14–15.

CHAPTER 9

Inter-conversion of embryonic limbs, see news report by Vogel, G. (1999) *Science*, 283: 1615–1616.

GENERAL

The accuracy of some old quotations was verified in Adams, F.P. (1952) *FPA's Book of Quotations*, New York: Funk and Wagnall.

INDEX

ablation 167

adenosine triphosphate *see* ATP

ADP (adenosine diphosphate) 100

aerobic life 33, 34–35

aluminium 14

amino acids 58–61, 68, 94, 139

   artificial generation of 118–119

ammonia, atmospheric 122–123

anaerobes 102–103

anatomy 30–22, 35

ancestry

   dating of 55–56

   human 35–37, 218

   of humans and apes 37–40

Anders, Edward 119, 126

Anders, William 2

Antarctic stone ALH84001 172

antibiotics 94

Apollo spacecraft 2

archaea (archaebacteria) 143–144

Archaeopteryx lithographica 41–42, 46

Ardipithecus 223

argon 57

Arrhenius, Svante 156

Artemia 156

arthropod 49

asteroids, re-entry of 163–171

atmosphere 172–173

   experimental 118–119

   planetary, primordial 118

atomic bomb 84

ATP (adenosine triphosphate) 35, 91, 92, 95, 100, 190, 191

   synthesis 105

ATP synthase 101, 102, 140

Bacillus 156

bacteria 140–141

bases, DNA 60, 66

Becquerel, Antoine Henri 10, 11

Belemnitella americana 86

Berner, Robert 106

Big Bang 16–17, 22, 23, 25, 30, 178, 182

biochemistry 33–35, 184

bipedalism 42–47

birds, relationships between reptiles, dinosaurs and 41–48

Blank, Carrine 89

brain, human 212–219, 223

Brasier, Martin 89

Brocks, Jochen 89, 90, 106

Buffon, Comte de *see* Leclerc

Butler, Samuel 145

Caenorhabditis elegans 59, 169

carbohydrates 99, 190

carbon 12 (^{12}C) 85, 86

carbon 12 : carbon 13 (^{12}C : ^{13}C) ratio 86–88

carbon 13 (^{13}C) 85, 86, 88–90

   depletion 87, 106

carbon compounds 84–90

cell components 92

cell formation 125–126

cell membrane 95–97, 137–138

cell replication 145–146

cell wall 137

cells 59–60

   early 136–139

cellular life 193–194

cellulose 92

Chambers, Robert 54, 204
Chambers, William 54
chemiosmotic hypothesis of oxidative
    phosphorylation 101–102
Chesapeake Bay crater 175
Chicxulub crater, Mexico 175
chlorophyll 104, 188, 189, 191,
    192
chloroplasts 145
Chomsky, Noam 222
closed universe 26–27
Clostridium 156, 157
codon 60, 79
Colymbosathon ecplecticos 65
comets 160–163
    orbits 170–171
    re-entry 163–170
common ancestors
    human 36–37, 218
    of humans and apes 37–40
communications 220–221
Confuciusornis sanctus 41
cosmological constant 27
coupling factors, respiratory 101
craters 2, 173–175
Creation, timing of 3–4
crystallins 146–147
Curie (unit) 11
Curie, Jacques 9, 10–11
Curie, Marie 9, 10, 11
Curie, Pierre 10
cuttlefish 86
cyanobacteria (blue-green algae) 88,
    89–90, 122
cytochrome 64, 68, 69, 91, 99–100,
    104, 192
cytochrome oxidase 99, 104, 107, 190,
    191, 192, 193

Dalrymple, Brent 4
Daphnia 156
dark matter 27
Darwin, Charles 3, 6–7, 233
    On the Origin of Species 6, 53

Dauphas, Nicolas 171
Dickens, Charles 212
dinosaurs
    demise of 108
    relationships between birds,
        reptiles and 41–48
distance ladder, stars 23
DNA (deoxyribonucleic acid) 50, 52,
    92, 93, 128, 155–156, 198
    dates from 58–61
    identical 66
    sequencing 63
Doppler, Christian 21
Doppler effect 20, 21
dormancy 116, 156–157
doubling 149–154
Drosophila 206–207

Earth
    age of 2–9, 108, 194–195
    early atmosphere 118
Eddington, Sir Arthur 169
Einstein, Albert 22, 27, 101, 220
energy 33–35
    consumption, human 34
entropy 114
enzymes 107–108, 126–127
Eros, asteroid 161
Escherichia coli 207
ether wind 101
eukaryotic cell 144, 145
evolutionary tree 38, 39, 41, 221,
    222
exons 81, 198
extinct life 40–48
eye 18, 49, 146

fatty acids 126, 137, 138
feathers 45, 46–47
fibrinopeptides, divergence in 64
flagella 139–140
flat universe 26, 27
flight 42, 44–46, 203–206
flightless birds 47

fossils 41–42, 45, 65
  dating 55–56
  formation 57
Frank, Louis 171
Fraunhofer, Joseph von 19

Galileo, space probe 161
gametophyte 183
Garcia-Ruiz, Juan Marcel 89
Gaspra, asteroid 161
gene pool 184, 213–214
genes 59–61, 184
  duplication 142–143
  mutations in 66, 73–74
  split 78–84
genetic code 239
genome 59
  human 61, 217
geological periods 174
glucose 125
glycine, synthesis of 118
Goethe, Johann Wolfgang von 54
gravitational attraction 163–164, 244
ground effect, in flight 46
GTP 91, 95

haemoglobin 69, 77, 78, 83, 91, 92,
    104, 105, 107, 186–189, 192
  divergence 64
  introns in 80
  precursor 187
  reversibility 107
hafnium ratio 13
half-lives, radioactivity 11–12
Halley's comet 162
haplosome 135, 136, 138
heat-shock proteins 146
Heisenberg's Uncertainty Principle
  16
Helmont, Johann Baptista van
  116–117, 121
hexokinase 184
hibernation 116, 156–157
histones 64, 68

Homo sapiens, appearance of 223
hormones 147
Hoyle, Fred 16, 17, 162
Hubble orbiting telescope 21
Hubble's constant 24, 25–26, 195,
    196, 234–236
humans, shortcomings of
  anatomical 200–202
  behavioural 202
Huxley, T.H. 9
Hyakutake, comet 162
hydrogen sources for photosynthesis
  99
hydrogen sulphide 99, 191–192
Hyman, Libbie Henrietta 50

insects
  evolution 209–210
  vs vertebrates 49
instinct 215–216
insulin 147
intelligence 199–200, 209–210, 218,
    219, 221
interplanetary transport 170–173
interstellar transport 170–173
introns 80, 81–83, 186, 187
invertebrates 48–50
irreducible complexity 96, 103, 110,
  121

Joly, John 7, 8

Kelvin, Lord (Sir William Thomson)
  5, 6, 8
Kimura, Motoo 66
kinetic energy 112
Kuiper Belt 162–165

lactate dehydrogenase 147
Laing, Samuel 8, 9, 180
Lake Acraman crater, Australia 174
language 222–223
Leclerc, George Louis, Comte de
  Buffon 4, 54–55

Leonids 166
Levy, David 165
life, origin of 28–29, 50–52, 177–179
  dating of 108–109
light waves 17–28, 155
Lightfoot, John 3, 4
lightning, as source of energy 118,
  119
limbs, vertebrate 204–206
Löb, Walther 118
Longrich, Nick 45
lutetium ratio 13
Lyrids 166
lysosomes 60

Mars
  life on 173
  meteorites on 2, 173
Matese, John 175
messenger RNA 82, 93, 94, 128
metamorphosis 183, 184
meteorites 2, 14, 158, 169–170, 243
meteors (shooting stars) 157–160, 243
Methanococcus jannaschii 143
Milky Way 175
Miller, Stanley 118–119, 126, 127
mini-genes 187
Mitchell, Peter 100–101
mitochondria 60, 101, 102, 145, 191
Mojzsis, Stephen 87
molecular clock 67–69
Monahans meteorite, Texas 170
Moon
  microcraters 2
  rocks, age of 14
Morokweng crater, Africa 175
Moyle, Jennifer 100
mutant proteins 66
mutations 42–43, 62, 63, 72–74, 142,
  198, 212–213
  neutral 66–67, 197
Muybridge, Eadweard 43
myoglobin 69, 105
Myxococcus xanthus 147

NADP 190
Near Earth Asteroid Rendezvous
  spacecraft (Shoemaker) 161, 165
neodymium 12
neutral theory of evolution 66–67, 197
nitrogen fixation 122
Nodularia 122
nucleotides 60, 91, 93–94

oleic acids 126
Oort cloud 159, 162, 164, 171, 175
open universe 26, 27
Oscillatoria 122
Osler, Sir William 53
osmosis 125
oxidative phosphorylation 100,
  101–102, 191
oxygen 104–108
  atmospheric, history of 105–108
  binding 187–188, 190
  energy and 97–103
  intake 32–33
oxytocin 147
ozone layer 119

Pali meterorite, Rajastan 168
palmitic acids 126
panspermia 156, 162
parsec (unit) 24–25
PDB, marine carbonate standard 85–87
penicillin 101
permeability, cell membrane 96–97
Perseids 166
Phillips, John 6
photosynthesis 90, 97–100, 102,
  103–104, 106, 190, 192, 193
Piltdown Man 40, 56, 237
plutonium 84
Poisson curve 72–74, 212, 240
Poisson formula 74–76
Popigai impact structure, Siberia 175
population growth 36
porphyrin 104
potassium 57

pre-cellular life 134–136
primordial life 90, 93–95
primordial soup 120–121, 123–126,
    134–135, 137
Primula kewensis 181
prokaryotes 139–145
Protarchaeopterix robusta 41
protein
    dates from 58–61
    divergence, measuring 72–78
    mutation rates 64–65, 66, 69
    origin of 127–128
    statistical chance of origins
        128–132
protein hormones 58
Protein S 147
protein sequencing 62–63, 185–186
    dating from 63–65

quadrupeds 43, 203–204

radio waves 155, 245
radioactive cosmochronology 28
radioactive decay 9
radioactivity, discovery of 9–13
radiometric dating 12
Reade, T. Mellard 7
red shift 21–24, 26, 27
re-entry, atmospheric 163–170
reflexes 216
reptiles, relationships between birds,
    dinosaurs and 41–48
respiration 100, 102–103
    anaerobes and 102–103
    origins of 103–104
    plant 99
ribosome 59, 93, 94–95, 103, 128,
    135, 136
Rivularia 122
RNA (ribonucleic acid) 61, 81, 82, 92,
    95
    messenger 82, 93, 94, 128
    transfer 94
RNA world 135–136

rocks, dating of 12–13, 56–57,
    87–89
Roentgen, Wilhelm 10
Rosing, Minik 87
rubidium 12, 57
rubisco (ribulose bis-phosphate
    carboxylase-oxygenase) 86, 91,
    92, 93, 99, 100
Rutherford, Ernest (Lord Rutherford
    of Nelson) 9

samarium 12
Samuel, Eugenie 173
Schiaparelli, Giovanni 166
Schopf, William 88, 89
selenium compounds 120
sexual reproduction 141, 142–143
Shapley, Harlow 23
Shipman, Pat 46
Shoemaker, Carolyn 165
Shoemaker, Eugene 165
Shoemaker–Levy 9, comet 165,
    166–167
single-celled organisms 50–51
skeleton 49
Slipher, Vesto 21
solar system, age of 13–16, 171
spectra 18–21, 28
split genes 78–84
sporophyte 183
starch 125
Stardust, spaceprobe 163
Steady State theory 17
stratigraphy 6
strontium 12, 57
Stubbs, George 43
Sun
    age of 5–6
    history and future of 8–9
Swan-Leavitt, Henrietta 23
Swift–Tuttle comet 159
Synechococcus 140
Système Internationale (SI) unit
    11

Taglish Lake meteorite, Canada 170
thermodynamics, laws of 111–112
    first 112–113, 133, 188
    second 62, 106, 112, 113–116, 124, 194
    third 116
thorium 28
Trans Neptunian Objects 162
transfer RNA 94
transposable exons 83
trexons 83
triangulation 23
Tunguska, comet impact on 166
Tyrannosaurus rex 43

Uncertainty Principle, Heisenberg 16
universe, age of 16–17, 179–182, 195
uranium 10–12, 28, 57, 84
Urey, Harold Clayton 118
Ussher, James 3, 4, 5, 54, 195

vacuum of space 17
vasopressin 147
vertebrates 35, 200
    vs insects 48–49
    uniformity and differences 30–32

viruses 162
volanic lava 119

Wallace, Alfred Russel 53
water 194
    on Earth 14–15
    from respiration 34–35, 99, 100
    as source of hydrogen 99
wavelength
    colour and 19–21
    of light, units 236–237
white dwarf 23
Whiting, Michael 32
Whitmire, Daniel 175
Wickramasinghe, Chandra 162
Wild 2, comet 163
wings 45–46, 47, 49
    genetic engineering of limbs 204–206
    human desire for 203–204
Wollaston, William 19
worms 49

X-rays, discovery of 10

zircons 15
Zwicky, Franz 22